Reading, Writing, and Doing

MATHEMATICAL PROOFS

Proof Techniques for

GEOMETRY

Daniel Solow

DALE SEYMOUR PUBLICATIONS

Printed in the United States of America. Published simultaneously in Canada.

DALE
SEYMOUR
PUBLICATIONS
P.O. BOX 10888
PALO ALTO, CA 94303

bcdefghi-MA-8987
Order Number DS01416
ISBN 0-86651-148-2

To Audrey, my wife and partner

CONTENTS

PREFACE TO THE INSTRUCTOR

The inability to communicate proofs in an understandable manner has plagued students and teachers in all branches of mathematics. The result has been frustrated students and teachers, and a contempt for theoretical mathematics.

One might conjecture that most students simply cannot understand abstract mathematics, but my experience indicates otherwise. What seems to have been lacking is a proper method for explaining theoretical mathematics. In this book, I have developed a systematic and organized method for communicating proofs; a common language that can be taught by professors and understood by students. In essence, this book categorizes, identifies, and explains (at the student's level) the various techniques that are used repeatedly in virtually all proofs.

Once students understand the techniques, it is then possible to explain any proof as a sequence of applications of these techniques. Explaining a proof in terms of its component techniques is not difficult, as is illustrated in the examples of this book. Before each "condensed" proof is a detailed analysis of the proof explaining the methodology, thought process, and techniques that are being used. Teaching proofs in this manner requires nothing more than preceding each step of the proof with an indication of which technique is about to be used, and why.

Simply stated, the objectives of this book are:

1. To enable a student to read proofs as might be found in a typical textbook or other mathematical literature. This goal can be accomplished by teaching the student to identify the techniques that are being used in the proof and how they are being applied to the specific problem.

2. To improve the problem-solving skills of students by teaching them to choose a proof technique consciously based on certain key words that appear in the problem.

While this book is not meant to teach creativity, I do believe that the proof techniques provide the tools needed to express creativity.

How to Use This Book

This book is designed for use as a workbook supplement for high school geometry courses. It explains the standard proof techniques that are encountered in typical geometry problems. Particular emphasis should be given to the Forward-Backward method of Chapter 2, since every other proof technique will be explained in terms of this method.

Since this book does not purport to teach geometry, it must be used in conjunction with another geometry text. Introduce the proof-techniques material in this book immediately after the students have been taught the following concepts:

1. Angles -- right, acute, obtuse, and measures of angles.
2. Polygons -- triangles (right, isosceles, equilateral, etc.); quadrilaterals (rectangles, squares, etc.).
3. Perpendicularity.

As you progress through this book, more and more concepts from geometry will be assumed. For example, the introductory proofs in Chapter 2 assume that the students have been exposed to the concepts of complementary and supplementary angles. Some of the contradiction proofs in Chapter 4 rely on the notion of parallel lines. The final two proofs in Chapter 8 deal with congruent triangles.

Each section of this book contains a short lesson and follow-up exercises. In the exercises, the students are asked to supply, with each proof, an analysis of the proof. The analysis requires students to show that they know what proof techniques they are going to use and why. In essence, the analysis should be thought of as an outline of the proof that will enable you to follow their thought processes and to identify incorrectly applied techniques. These analyses should be similar in nature to the ones found in the body of the text, although perhaps more condensed. The answers to the exercises provide numerous examples to guide you. It is beneficial to correct the students when they write their very first analysis so that their remaining ones are then prepared properly.

Since eleventh and twelfth graders (who have already learned geometry) might also be working through the first part of this book outside of a geometry course, there are numerous starred sections(*) and exercises designed exclusively for their use.

ACKNOWLEDGEMENTS

With regard to the preparation of this book, I am deeply indebted to my student, Carolyn Cuff, who made many substantive suggestions and spent a great deal of time proofreading and correcting the manuscript. I thank Diane Bellis for taking the time to carefully work through all of the exercises. Technical editing comments by Elaine Murphy were also appreciated. I am grateful to my wife, Audrey, for the long hours spent in data-basing, editing, and proofreading the book. I also thank Evelyn Bell and Dale Seymour for their constructive comments and for their timely support of this work.

Finally, I am grateful to Hosei University in Tokyo where, as a visiting researcher, I was able to complete much of this work.

PROOF TECHNIQUES FOR GEOMETRY

CHAPTER 1
The Truth of It All

1.1 WHAT IS A PROOF?

The objective of mathematicians is to discover and to communicate certain truths. Mathematics is the language of mathematicians, and a proof is a method of communicating a mathematical truth to another person who also "speaks" the language. This book explains much of the basic "grammar" that you will need in order to learn the language of mathematics.

One of the major reasons for doing proofs is to establish, beyond any reasonable doubt, that certain mathematical statements are true. In general, mathematical statements can be either true or false. Some examples of statements would be:

1. A triangle has three sides.
2. $1 = 0$.
3. $3x = 5$ and $y = 1$.
4. x is not greater than 0.
5. The sum of the degrees of the three angles in any triangle is 180°.

Observe that statement (1) is always true, statement (2) is always false, and statements (3) and (4) can be either true or false, depending on the value of a variable.

It is perhaps not as obvious that statement (5) is always true. It therefore becomes necessary to have some method for proving that such statements are true. In other words, a mathematical proof is a convincing argument that is expressed in the language of mathematics. Thus, a proof should contain enough mathematical detail so as to be convincing to the person(s) to whom the proof is addressed. For instance, a proof of statement (5) written by a mathematics teacher aimed at convincing another mathematics teacher might consist of nothing more than Figure 1. On the other hand, a proof directed toward a geometry student would probably require greater detail, perhaps even the definition of a triangle. All of your proofs should be aimed at convincing another student in your class. More will be said about writing proofs in the next chapter.

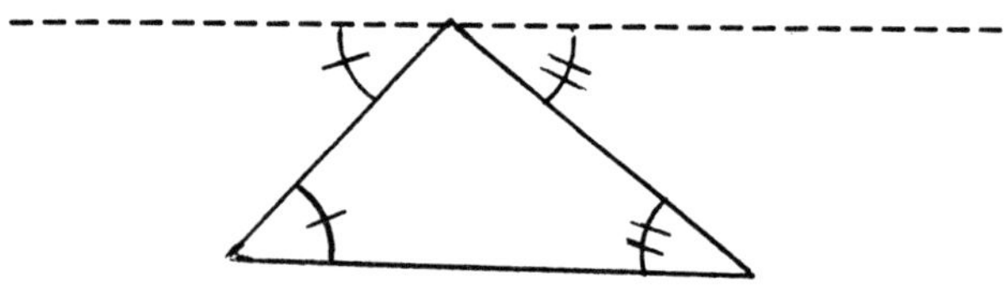

Figure 1. A Proof That a Triangle Has 180°.

Just as music is composed of a sequence of individual notes, so proofs consist of individual proof techniques. This book will describe the fundamental techniques. Then, you will see how all proofs can be understood as a sequence of applications of the individual techniques. Unfortunately, most proofs in textbooks make little or no reference to which techniques are being used. One objective of this book is to teach you how to read these "condensed" proofs. More will be said about reading proofs in the next chapter.

The final objective of this book is to improve your problem-solving skills by showing you that, often, the very form of the problem under consideration can be used to choose a correct technique with which to do the proof. You will be taught that when a certain key word appears in your problem, an associated proof technique is likely to be successful in solving the problem. Thus, when attempting to create your own proofs, learn to select a proof technique consciously before wasting hours trying to figure out what to do.

By learning the thought process involved in proofs, you will be able

1. To establish, beyond any reasonable doubt, that certain mathematical statements are true.
2. To communicate to others precisely and unambiguously why such statements are true.
3. To read and to understand almost any mathematical subject on your own.
4. To approach and to solve many problems (even non-mathematical ones) in a systematic and organized manner.

EXERCISES FOR SECTION 1.1

Fill in the blanks with one or more words.

1. A mathematical statement must be either ____________ or ______________.

2. A proof is a ________________________ that is expressed in the language of mathematics.

3. For you, a proof should be aimed at convincing _________ ___________________________.

4. A proof can be broken down into a sequence of individual ____________________________.

5. A particular proof technique should be chosen based on certain ______________________ that appear in the problem under consideration.

Indicate which of the following are valid mathematical statements by circling the appropriate ones. Remember, a valid mathematical statement must be either true or false.

6. $ax^2 + bx + c = 0$.

7. $-b \pm \sqrt{b^2 - 4ac}$.

8. Triangle XYZ is isosceles.

9. $3 + n + n^2$.

10. Every quadratic equation has a solution that is a real number.

11. n^2 is an even integer.

12. Acute angles and right angles.

1.2 IF ... THEN ... STATEMENTS

Mathematical proofs can be applied to the following types of problems. Given two statements A and B, establish that

If A is true then B is true.

For example, a proof could be applied to show that

If the sun is shining then it is hot.

In this example one has

Statement A: The sun is shining.
Statement B: It is hot.

These "if ... then ..." statements are called <u>conditionals</u>, or <u>implications</u>. Statement A is called the <u>hypothesis</u> and B the <u>conclusion</u>. In general, everything after the word "if" and before the word "then" comprises the hypothesis A. Everything after the word "then" constitutes the conclusion B. Here are some examples of conditionals.

Example: If x is a real number less than 1, then $x^2 + 3x - 2 < 2$.
Hypothesis A: x is a real number less than 1.
Conclusion B: $x^2 + 3x - 2 < 2$.

Example: If two lines intersect, then the vertical angles are equal.
Hypothesis A: Two lines intersect.
Conclusion B: The vertical angles are equal.

Sometimes statements must be rewritten so that the hypothesis and conclusion can be identified more easily.

Example: In the equilateral triangle XYZ, all three angles are equal.

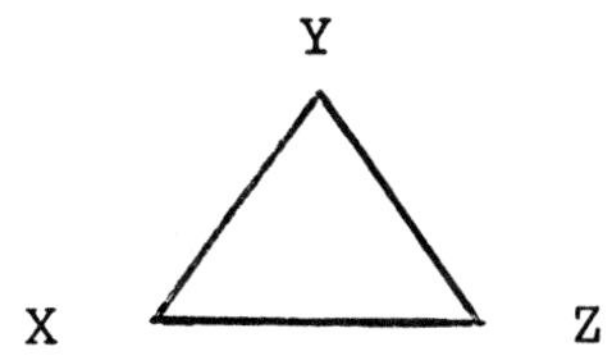

Rewritten: If triangle XYZ is equilateral, then $\angle X = \angle Y = \angle Z$. (Note: In this book, the symbol $\angle X$ will be used to mean either angle X or the number of degrees in angle X, depending on the context in which the symbol is used.)

Hypothesis A: Triangle XYZ is equilateral.

Conclusion B: $\angle X = \angle Y = \angle Z$.

Mathematicians are often very lazy. They have shortened the statement

If A is true then B is true

to

If A then B

or simply

A implies B.

When it comes to writing, mathematicians have developed a symbolic "shorthand" notation. For instance, the statement "A implies B" would be written

$A \Rightarrow B$.

This book will include the appropriate symbols but will not use them in proofs. A complete list of the symbols can be found in the glossary at the end of this book.

Truth of It All

A proof is a formal method for showing that a conditional statement is true. So, in order to do a proof, you must know exactly what it means to show that

If A is true then B is true.

Specifically, doing a proof consists of

1. Assuming that statement A is true.
2. Using the assumption that A is true to reach the conclusion that B is true.

The proof techniques described in this book are designed to help you to accomplish (2).

Since you will be assuming that statement A is true in your proof, the information contained in A is often referred to as the given information, or simply, the "given." Similarly, statement B is sometimes referred to as the "to prove" statement.

It is important to note that when you are doing a proof, you are not trying to show that A is true. Nor are you trying to determine if B is true. Rather, you are trying to show that B is true as a logical consequence of assuming that A is true. Thus, your ability to show that B is true will depend heavily on the fact that you have assumed A to be true and, ultimately, you will have to discover the linking relationship between A and B. Doing so will require a certain amount of creativity on your part. The proof techniques in this book are designed to get you started and to guide you along the path.

Hereafter, A and B will be statements that are either true or false. The problem of interest will be to show that

If A is true then B is true.

EXERCISES FOR SECTION 1.2

For each of the following conditionals, write down the hypothesis and the conclusion.

1. If n is an even integer, then n^2 is an even integer.

   Hypothesis: ______________________________

   Conclusion: ______________________________

2. ∠ 1 and ∠ 2 are right angles implies ∠ 1 and ∠ 2 are supplementary.

   Hypothesis: ______________________________

   Conclusion: ______________________________

3. The sum of the first n positive integers is $n(n + 1)/2$.

   Hypothesis: ______________________________

   Conclusion: ______________________________

For each of the following problems, indicate what you would assume and what you would conclude when doing a proof.

4. n is an integer and n^2 is even => n is an even integer.

   Assume: ______________________________

   Conclude: ______________________________

5. If p and q are positive real numbers with $\sqrt{pq} \neq (p + q)/2$, then $p \neq q$.

   Assume: ______________________________

   Conclude: ______________________________

6. When x is a real number, the minimum value of $x(x - 1)$ is at least $-1/4$.

   Assume: ______________________________

   Conclude: ______________________________

For each of the following problems, indicate the "given" and "to prove" statements.

7. For every real number $x > 1$, $x > 1/x$.

   Given: ______________________________

   To Prove: ______________________________

8. Two integers whose sum is odd cannot both be even.

   Given: ______________________________

   To Prove: ______________________________

9. For a real number $x \neq 0$, it follows that $|x| > 0$.

   Given: ______________________________

   To Prove: ______________________________

10. Circle the one statement that is different from the others.

    a. A implies B.

    b. If A is true then B is true.

    c. A => B.

    d. If B is true then A is true.

    e. If A then B.

*1.3 TRUTH TABLES

There is another useful way of thinking about the statement "A implies B." Just like any other statement, this one can be either true or false. It seems reasonable that the conditions under which the statement "A implies B" is true will depend on whether A and B themselves are true. Consequently, there are four cases to consider:

1. A is true and B is true.
2. A is true and B is false.
3. A is false and B is true.
4. A is false and B is false.

Suppose, for example, that your friend made the statement:

If it rains then Mary brings her umbrella.

Here one has the statements:

A: It rains.
B: Mary brings her umbrella.

To determine when the statement "A implies B" is false, ask yourself in which of the four cases would you be willing to call your friend a liar. In the first case, it does rain and Mary does bring her umbrella, so your friend has told the truth. In the second case, it has rained and yet Mary did not bring her umbrella, as your friend said she would. Here your friend has not told the truth. In the third and fourth cases, it does not rain. You would not really want to call your friend a liar in the case of no rain because, your friend only said that something would happen if it did rain. Thus, the statement "A implies B" is true in each of the four cases except the second one, as is shown in Table 1.

Table 1. The Truth of "A implies B"

A	B	A implies B
True	True	True
True	False	False
False	True	True
False	False	True

Table 1 is an example of a truth table. A truth table is a method for determining when a complex statement (in this case, "A implies B") is true by examining all possible truth values of the individual statements (in this case, A and B). According to Table 1, when trying to show that "A implies B" is true, you can assume that the statement to the left of the word "implies" (namely A) is true. Your goal is to use that assumption in order to reach the conclusion that the statement to the right of the word "implies" (namely B) is true.

Another example of a truth table arises from the statement A AND B (written $A \wedge B$), which is true only when A is true and B is true. The associated truth table is given in Table 2.

Table 2. The Truth of A AND B

A	B	A AND B
True	True	True
True	False	False
False	True	False
False	False	False

A final example would be the truth table associated with the statement A OR B (written $A \vee B$), which is false only when A is false and B is false. The truth table is presented in Table 3.

Table 3. The Truth of A OR B

****

A	B	A OR B
True	True	True
True	False	True
False	True	True
False	False	False

****

EXERCISES FOR SECTION 1.3

Use Table 1 to determine whether each of the following statements are true or false, and give your reason.

1. If $2 > 7$ then $1 > 3$.

   ____

2. If $2 < 7$ then $1 < 3$.

   ____

3. If $x = 3$ then $1 < 2$.

   ____

4. If $x = 3$ then $1 > 2$.

   ____

Fill in the truth table for each of the following statements. (Note: T = True, F = False)

5. Statement: NOT A -- which is true when A is false.

A	NOT A
T	___
F	___

6. Statement: A implies (B implies C).

A	B	C	B => C	A => (B => C)
T	T	T	___	___
T	T	F	___	___
T	F	T	___	___
T	F	F	___	___
F	T	T	___	___
F	T	F	___	___
F	F	T	___	___
___	___	___	___	___

7. Statement: (A implies B) implies C.

A	B	C	A => B	(A => B) => C
___	___	___	___	___
___	___	___	___	___
___	___	___	___	___
___	___	___	___	___
___	___	___	___	___
___	___	___	___	___
___	___	___	___	___
___	___	___	___	___

8. If you are trying to show that "A implies B" is true, and you know that B is false, do you want to show that A is true or that A is false? Explain.

____

____

CHAPTER 2
The Forward-Backward Method

2.1 WHAT IS THE FORWARD-BACKWARD METHOD?

As you learned in the previous chapter, much of mathematics is concerned with starting from some given information contained in the hypothesis A, and trying to reach a desired conclusion B. The forward-backward method is the fundamental proof technique for accomplishing that goal.

To see how the forward-backward method works, try to solve the maze in Figure 2. Most likely, you began by putting your finger on the point A and proceeded in the forward direction until you were blocked. Then you probably tried a different path and, if you persisted, you finally reached the point B.

When you got stuck in going forward, did it ever occur to you to work backward from B? Specifically, you could put your finger on the point B and "back up" along any path that leads to B. Stopping at some point, say B1, you should say to yourself, "If I could somehow reach B1, then I could also get to B." In fact, if you can back up all the way to where you stopped going forward, then you will have solved the problem!

In mathematics, you work forward by starting with the information in the hypothesis A. Your objective is to reach the desired conclusion B. If you get stuck in the forward process, then the backward process is available to help you back up from B.

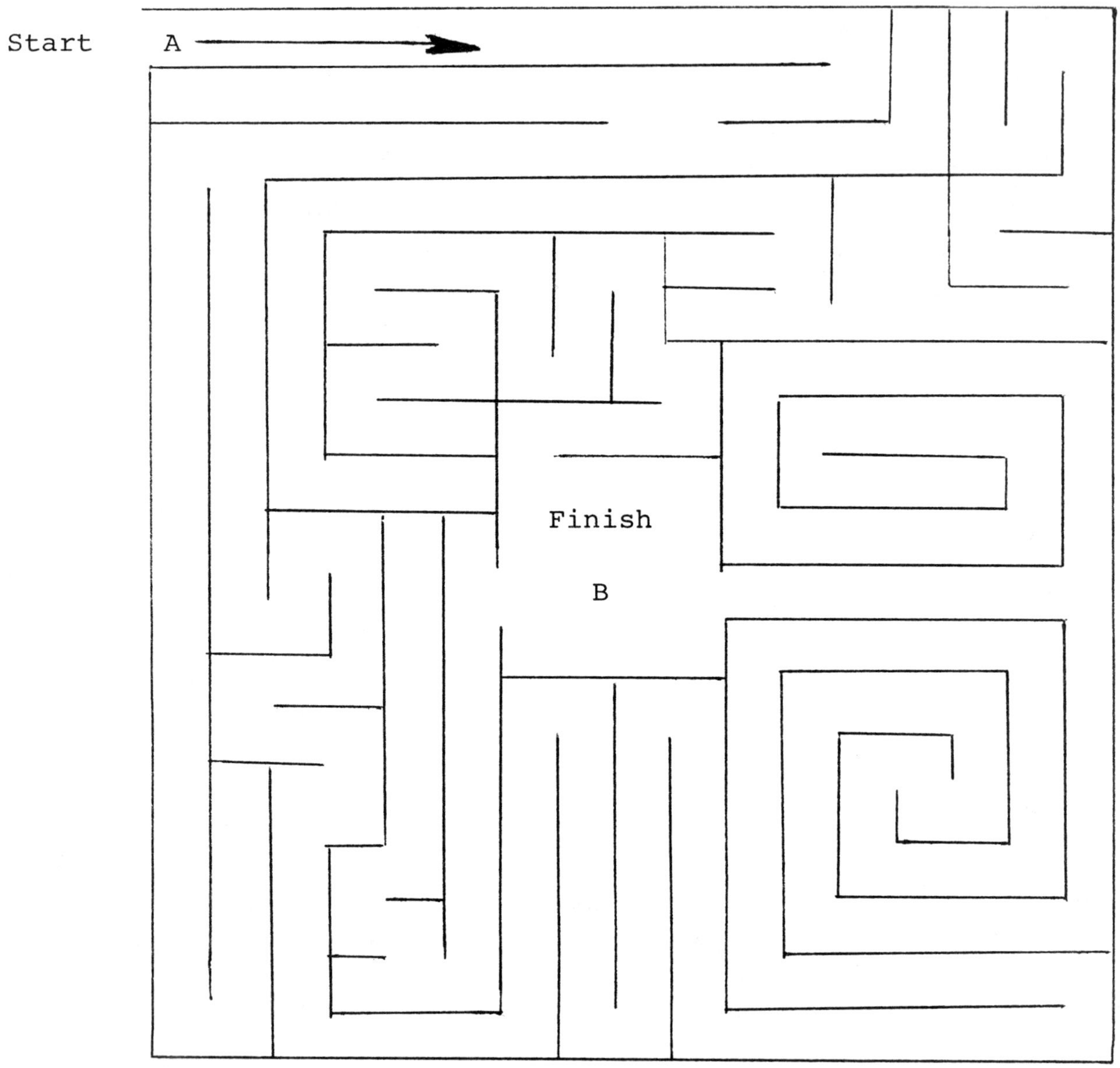

Figure 2. The Maze.

The Forward Process

To work forward in mathematical problems, you start with statement A, and derive from it other statements that are true as a result of assuming that A is true. For example, if you had the statement

A: $5x - 10 = 0$

then, by adding 10 to both sides of the equality, a new statement in the forward process would be

A1: $5x = 10$.

There are many ways to work forward from A. For instance, you could have divided both sides of the equality in A by 5 to obtain

A1: $x - 2 = 0$.

The forward process could then be applied to either of the two statements A1 so as to generate other true statements. Always keep in mind that the forward process is directed toward producing statement B. When you are unable to work forward for some reason, it is time to look at B and work backward.

The Backward Process

To use the backward process in solving mathematical problems, you look at the conclusion B, and try to derive from it a new statement, say B1, with the property that if B1 were true, then B would be true. For example, if the desired conclusion of a problem were

B: $x = 2$

then one step in the backward process might lead to the statement

B1: $x - 2 = 0$ (which also happens to be A1).

Looking at B1, you should say to yourself, "If B1 were true, then B would be true."

You could then try to back up from B1. Just keep in mind that the objective of the backward process is to reach the last statement in the forward process.

EXERCISES FOR SECTION 2.1

Fill in the blanks with one or more words.

1. In the forward process, you start with the ____________ and use it to produce other true statements.

2. In the backward process, you obtain from the ________________ a new statement such that if it is true, then so is the _______________.

3. The objective of the forward process is to reach the _______________.

4. The objective of the backward process is to reach the ______________.

5. When you cannot go forward in a constructive way, you should ____ ____.

6. If you start with the backward process and get stuck, you should ____ ____.

2.2 USING THE FORWARD-BACKWARD METHOD FOR SOLVING PROBLEMS

The forward-backward method will now be used to solve a puzzle. On the planet Ork, there is an alphabet consisting of the letters s and t together with the following rules for creating new words from old ones.

Rule 1. Double the word.
Example: The word sts would become stssts.

Rule 2. Erase tt from the word.
Example: The word stts would become ss.

Rule 3. Replace sss in the word by t.
Example: The word stsss would become stt.

Rule 4. Add the letter t to the end of the word if its last letter is s.
Example: The word tss would become tsst.

The Problem

Starting with the word s, apply the rules above, in any order and as often as you like, to obtain the word sst.

The Solution

The Forward Process

First, use the forward process by applying each rule exactly once to the word s in an attempt to produce the desired word sst. Thus, from the word s you could obtain

By rule 1: ss.

By rule 2: Not applicable.

By rule 3: Not applicable.

By rule 4: st.

It is not clear whether to proceed forward from ss or from st. To find out, turn to the backward process.

<u>The Backward Process</u>

To use the backward process, look at the desired conclusion sst, and ask yourself, "What word could I start with that would then lead to sst?" Specifically, write down a list of all words for which an application of one of the above rules would result in sst.

Rule 1: No word could be doubled and result in sst.

Rule 2: ssttt (the last two t's could be erased to obtain sst), or

sttst (the middle two t's could be erased to obtain sst), or

ttsst (the first two t's could be erased to obtain sst).

Rule 3: sssss (the last three s's could be replaced by t to obtain sst).

Rule 4: ss (the letter t could be added to obtain sst).

Looking at the word ss from rule 4 and at the results of the forward process, perhaps now it is clear how to solve the problem. Start with s, apply rule 1 to obtain ss, then apply rule 4 to obtain sst, that is,

1. s (given).

2. ss (rule 1).

3. sst (rule 4).

EXERCISES FOR SECTION 2.2

Fill in the blanks with the appropriate words.

1. Working forward by applying rule 3 exactly once to the word ssss would result in either ________________ or ______________.

2. Working backward, you could obtain the word tst by applying rule 4 exactly once to the word ______________.

3. Combining the forward and backward steps of Exercises 1 and 2, you can start with the word ssss and obtain tst by first applying rule _______ to ssss obtaining ______, and then applying rule _______ to obtain tst.

Apply the forward process exactly once to the given word using the indicated rules. If a rule cannot be applied, write "NA" in the appropriate space.

4. ss: ________ rule 1 ________ rule 2 ________ rule 3 ________ rule 4

5. stt: ________ rule 1 ________ rule 2 ________ rule 3 ________ rule 4

6. ttsss: ________ rule 1 ________ rule 2 ________ rule 3 ________ rule 4

Forward-Backward

Work backward one step from the given word by writing down <u>all</u> words for which the indicated rule would result in the given word. If no word can be written down for a specific rule, write "NA" in the appropriate space.

Example: ts:	NA	ttts or tstt	ssss	NA
	rule 1	rule 2	rule 3	rule 4

7. t:	_______	_______	_______	_______
	rule 1	rule 2	rule 3	rule 4
8. tt:	_______	_______	_______	_______
	rule 1	rule 2	rule 3	rule 4
9. st:	_______	_______	_______	_______
	rule 1	rule 2	rule 3	rule 4

Use the results of your practice in Exercises 4 through 9 to fill in the following blanks and to solve the following problems.

10. To obtain the word st from the word ss, start with ss and apply rule _______ to obtain the word _________ then apply rule ________ to obtain the word st.

11. To obtain the word st from the word stt, start with stt and apply rule ______ to obtain the word _____, then apply rule _____ to obtain the word st.

12. To obtain the word t from the word ttsss, start with ttsss and apply rule _____ to obtain the word _____, then apply rule _____ to obtain the word t.

For each word listed below, indicate which rule or rules were used to obtain that word from the word immediately above it.

13.

s	given	
ss	rule	_______
ssss	rule	_______
sssst	rule	_______
tst	rule	_______

14.

tst	given			
tsst	rule	_______	then rule	_______
tssttsst	rule	_______		
ttst	rule	_______	then rule	_______

2.3 USING THE FORWARD-BACKWARD METHOD FOR SOLVING MATHEMATICAL PROBLEMS

In the following example, recall that a right angle has 90 degrees, and that supplementary angles are angles whose degrees add up to 180°.

Example 1: If $\angle$ RST and $\angle$ RSU are right angles, then $\angle$ RST and $\angle$ RSU are supplementary.

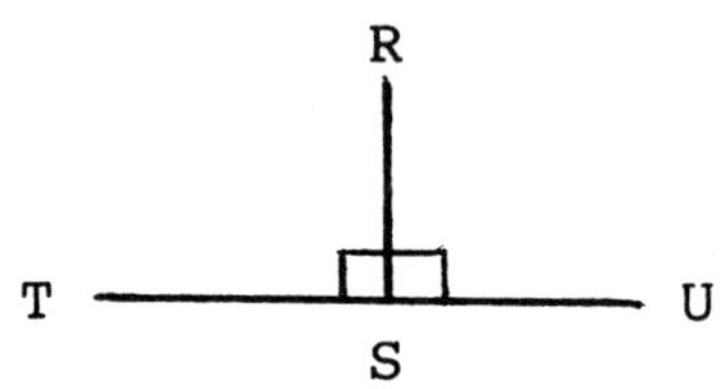

Analysis: First identify the hypothesis A and the conclusion B, so

A: $\angle$ RST and $\angle$ RSU are right angles.

B: $\angle$ RST and $\angle$ RSU are supplementary.

<u>The Backward Process</u>

Look at statement B and ask yourself how or when you could reach the conclusion. The specific question you ask yourself will be called the <u>key question</u>. For this problem, the correct key question to ask is the

Key Question: How can I show that two angles are supplementary?

Note: A properly posed key question should contain no symbols or other notation from the specific problem.

The next step in the backward process is to answer the key question. Can you do so for the key question above? Certainly one answer is the

Key Answer: Show that the sum of the degrees of the two angles is 180°.

Referring back to the figure, this means that you should show that

B1: $\angle RST + \angle RSU = 180^{\circ}$.

Observe that answering the key question is a two-step process.

Step 1. Answer the key question without using symbols or other notation from the specific problem.

Step 2. Apply the answer of Step 1 to the specific problem using the appropriate symbols and notation.

The net result of the backward process is a new statement, B1, with the property that if B1 is true, then so is B. Now, all of your efforts should be directed toward reaching the conclusion that B1 is true. To do so, one alternative is to work forward from the hypothesis A.

The Forward Process

To use the forward process, look at statement A and ask yourself if there is any statement that you can say is true as a result of A. Since right angles have 90 degrees, you can say that

A1: $\angle RST = 90^\circ$.

A2: $\angle RSU = 90^\circ$.

Recall that the forward process is directed toward producing precisely the last statement that was obtained in the backward process, namely

B1: $\angle RST + \angle RSU = 180^\circ$.

You can obtain B1 by adding both sides of the equalities in A1 and A2, thus completing the proof.

Summary of this Proof

The steps of this proof will now be summarized. More will be said about writing and reading proofs later.

PROOF

STATEMENT	REASON
1. $\angle RST$ and $\angle RSU$ are right angles.	1. Given in the hypothesis.
2. $\angle RST = 90^\circ$.	2. Definition of right angle.
3. $\angle RSU = 90^\circ$.	3. Definition of right angle.
4. $\angle RST + \angle RSU = 180^\circ$.	4. Adding (2) and (3).
5. $\angle RST$ and $\angle RSU$ are supplementary.	5. Definition of supplementary angles.

EXERCISES FOR SECTION 2.3

Supply a key question for each of the following desired conclusions. Be sure that your key question uses no symbols from the problem.

1. Desired Conclusion: $\angle$ RST is a right angle.

   Key Question: How can I show that ____________________

   ____?

2. Desired Conclusion: Triangle XYZ is isosceles.

   Key Question: How can I show that ____________________

   ____?

3. Desired Conclusion: Lines R and S are perpendicular.

   Key Question: How can I show that ____________________

   ____?

4. Desired Conclusion: $\angle$ 1 and $\angle$ 2 are complementary.

   Key Question: How can I show that ____________________

   ____?

5. Desired Conclusion: $\{\text{numbers } x\text{: } x^2 \leq 1\} =$

   $\{\text{numbers } x\text{: } |x| \leq 1\}$.

   Key Question: How can I show that ____________________

   ____?

For each of the following conclusions B and given key questions, use the associated figure to write a key answer that uses no symbols or notation from the problem. Then, apply your key answer to the specific problem obtaining a statement B1 that does use the notation of the problem.

6. 

Conclusion B: $\angle 1$ and $\angle 2$ are supplementary.

Key Question: How can I show that two angles are supplementary?

Key Answer: Show that ______________________________

____

B1: ______________________________________

7. 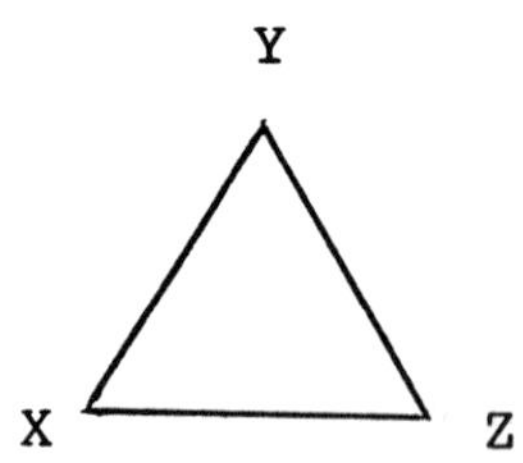

Conclusion B: Triangle XYZ is equilateral.

Key Question: How can I show that a triangle is equilateral?

Key Answer: Show that ________________________________

____

B1: ____

8.

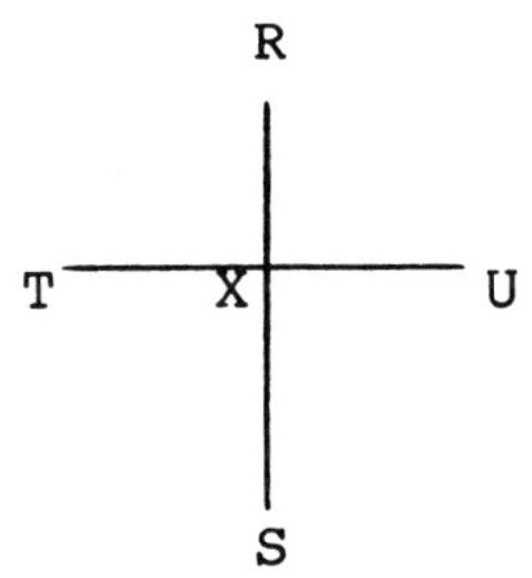

Conclusion B: Lines RS and TU are perpendicular.

Key Question: How can I show that two lines are perpendicular?

Key Answer: Show that ________________________________

____

B1: ____

For each of the given statements A, use the forward process one step to derive three different statements that are true.

9. A: The real number x satisfies $x^2 - 3x + 2 \leq 0$.

A1: ________ A1: ________ A1: ________

10.

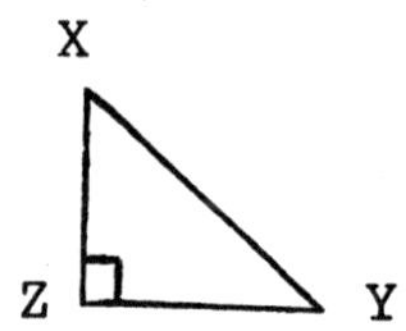

A: The triangle XYZ is an isosceles right triangle.

A1: ________ A1: ________ A1: ________

11.

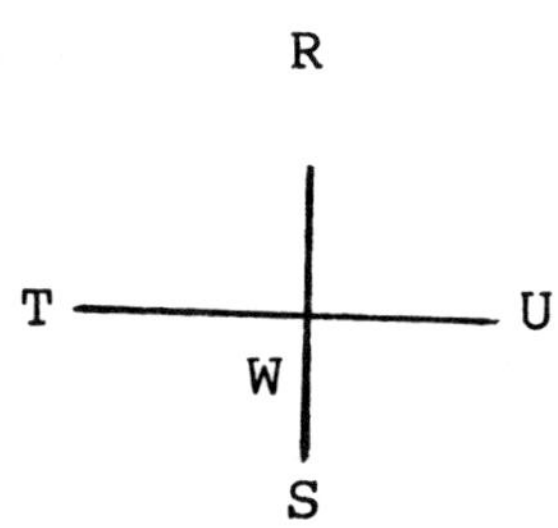

A: The line segments RS and TU are perpendicular, and they bisect each other.

A1: ________ A1: ________ A1: ________

*12. For the conclusion of the problem "If x is a real number, then the maximum value of $-x^2 + 2x + 1$ is greater than or equal to 2," circle the incorrect key question.

(a) How can I show that the maximum value of a parabola is greater than equal to a number?

(b) How can I show that a number is less than or equal to the maximum value of a polynomial?

(c) How can I show that the maximum value of the function $-x^2 + 2x + 1$ is greater than or equal to a number?

(d) How can I show that a number is less than or equal to the maximum of a quadratic function?

What is incorrect about the key question you circled?

____

____

13. For the conclusion of the problem "If

$$R = \{\text{real numbers } x: \; x^2 - x \leq 0\}$$

$$S = \{\text{real numbers } x: \; -(x - 1)(x - 3) \geq 0\}$$

$$T = \{\text{real numbers } x: \; x \geq 1\}$$

then the intersection of R and S is a subset of T," circle the correct key question.

(a) How can I show that a set is a subset of another set?

(b) How can I show that the intersection of R and S is a subset of T?

(c) How can I show that every point in the intersection of R and S is greater than or equal to 1?

(d) How can I show that the intersection of two sets has a point in common with another set?

Briefly explain what is wrong with the other choices.

(a) ____________________

(b) ____________________

(c) ____________________

(d) ____________________

2.4 ANOTHER EXAMPLE FROM MATHEMATICS

In the following example, recall that two angles are complementary if the sum of their degrees is 90°. Also, two line segments are perpendicular (written $\perp$) if the angles formed at their intersection are right angles.

Example 2: If the line segments RT and ST are perpendicular, and $\angle 1 + \angle 2 + \angle 3 = 180^\circ$, then $\angle 1$ and $\angle 2$ are complementary.

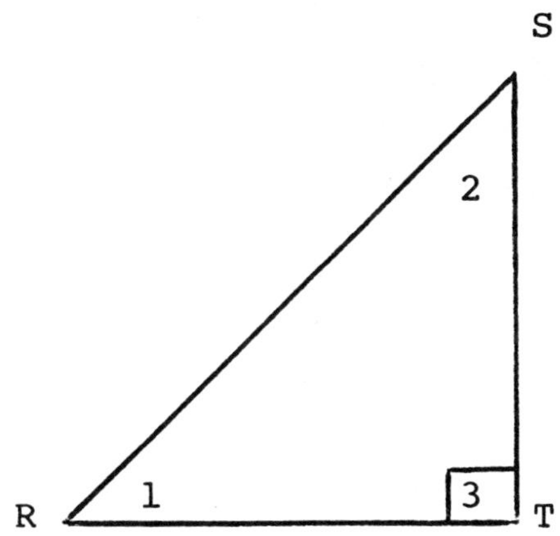

Analysis: First identify the hypothesis A and the conclusion B, so

A: RT and ST are perpendicular, and $\angle 1 + \angle 2 + \angle 3 = 180^\circ$.

B: $\angle 1$ and $\angle 2$ are complementary.

<u>The Backward Process</u>

Looking at statement B, you should ask the

Key Question: How can I show that two angles are complementary?

One possible answer is the

Key Answer: Show that the sum of the degrees of the two angles is 90°.

Applying the key answer to the specific problem, you obtain the new statement to be proved:

B1: $\angle 1 + \angle 2 = 90^{\circ}$.

All efforts should now be directed toward reaching the conclusion that B1 is true. To do so, you could work forward from the hypothesis A; however, to illustrate some of the difficulties that can arise in the backward process, let's continue working backward from B1. Can you pose a key question for B1? Certainly one could ask the

Key Question: How can I show that the degrees of two angles add up to 90°?

One could also ask the equally valid

Key Question: How can I show that two numbers are equal?

After all, the quantity $\angle 1 + \angle 2$ is a number, and so is 90°. Thus, there can be more than one key question.

Also, a particular key question can have many key answers, and you might not know which one to choose. For instance, associated with the last key question above could be either of the following:

Key Answer: Show that the difference of the two numbers is 0.

Key Answer: Show that each of the two numbers is equal to some other third number.

When you encounter these complications in the backward process, try to use the information in the hypothesis A to guide you. For example, the hypothesis might help you to choose a particular key question when there is more than one. If, however, you are still unable to proceed with the backward process, then turn to the forward process.

The Forward Process

The objective now is to work forward from statement A to obtain statement B1. Look at A, in particular, the fact that RT and ST are perpendicular. You can use your knowledge of perpendicularity to say that

A1: $\angle 3 = 90^\circ$.

Also, from statement A, you have

A2: $\angle 1 + \angle 2 + \angle 3 = 180^\circ$.

Can you see how to combine A1 and A2 to obtain B1? You need only subtract the equality in A1 from that in A2, obtaining precisely

B1: $\angle 1 + \angle 2 = 180^\circ - 90^\circ = 90^\circ$

thus completing the proof.

Summary of this Proof

The steps of this proof will now be summarized. Observe the complete lack of reference to the backward process and key questions.

PROOF

STATEMENT	REASON
1. RT $\perp$ ST.	1. Given in the hypothesis.
2. $\angle 3 = 90^\circ$.	2. Definition of perpendicularity.
3. $\angle 1 + \angle 2 + \angle 3 = 180^\circ$.	3. Given in the hypothesis.
4. $\angle 1 + \angle 2 = 90^\circ$.	4. Subtract (2) from (3).
5. $\angle 1$ and $\angle 2$ are complementary.	5. Definition of complementary angles.

EXERCISES FOR SECTION 2.4

For each of the following conclusions, list two different key questions that use no symbols or notation from the specific problem.

1. If n is an even integer, then n^2 is an even integer.

   Key Question: How can I show that ______________________

   ____?

   Key Question: How can I show that ______________________

   ____?

2. If, in the right triangle XYZ, $\angle X = \angle Y$, then YZ = XZ.

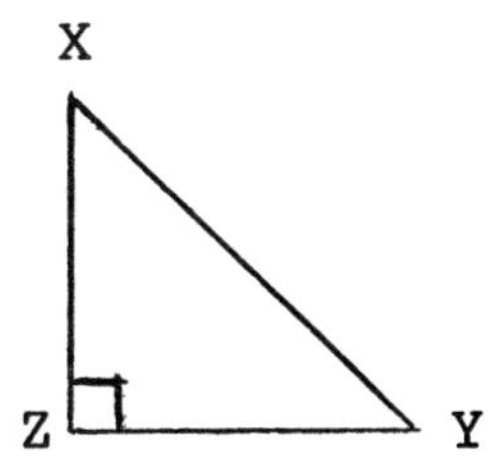

   Key Question: How can I show that ______________________

   ____?

   Key Question: How can I show that ______________________

   ____?

3. If n is a given integer that satisfies the equation $-3n^2 + 2n + 8 = 0$, then $2n^2 - 3n = 2$.

Key Question: How can I show that ______________________

____?

Key Question: How can I show that ______________________

____?

4. If R and S are sets, then S is a subset of R union S.

Key Question: How can I show that ______________________

____?

Key Question: How can I show that ______________________

____?

For each of the following key questions, provide two key answers.

5. Key Question: How can I show that two real numbers are equal?

(Note: Provide answers that are different from the ones given in the text.)

Key Answer: Show that ______________________

Key Answer: Show that ______________________

6. Key Question: How can I show that a triangle is a right triangle?

Key Answer: Show that ________________________________

Key Answer: Show that ________________________________

7. Key Question: How can I show that two angles are equal?

Key Answer: Show that ________________________________

Key Answer: Show that ________________________________

For each of the conclusions to the following problems, (1) pose a key question, (2) provide a key answer, and (3) apply the key answer to the specific problem to obtain a new statement B1 that is to be proved.

8. If $\angle R = \angle S = \angle T$, then triangle RST is equilateral.

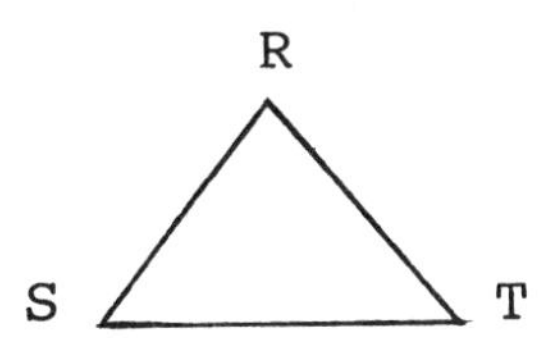

Key Question: How can I show that ______________________

____?

Key Answer: Show that ______________________________

____

B1: ____

9. If ∠ TSU = ∠ USV and RU is a line, then ∠ USV and ∠ RST are supplementary.

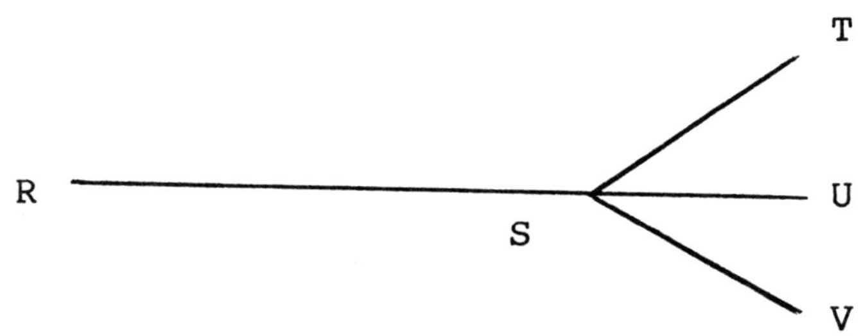

Key Question: How can I show that ______________________

____?

Key Answer: Show that ______________________________

____

B1: ____

For each of the following statements B, write down a key question. Then use statement B1 to supply the appropriate key answer.

10. B: $\angle 1 + \angle 2 = 90^\circ$.

B1: $\angle 1 + \angle 2 + 90^\circ = 180^\circ$.

Key Question: How can I show that ______________________

____?

Key Answer: Show that ______________________________

____

11. B: $x - y = 0$.

B1: $(x - y)^2 = 0$.

Key Question: How can I show that ______________________

____?

Key Answer: Show that ______________________________

____

*2.5 AN ADVANCED EXAMPLE OF THE FORWARD-BACKWARD METHOD

In the following example, recall that the area of a right triangle is one-half the base times the height, and that the Pythagorean theorem states that the square of the length of the hypotenuse equals the sum of the squares of the lengths of the other two sides. Also recall that an isosceles triangle is one that has two sides of equal length.

Example 3: If the right triangle XYZ has an area of $z^2/4$, then triangle XYZ is isosceles.

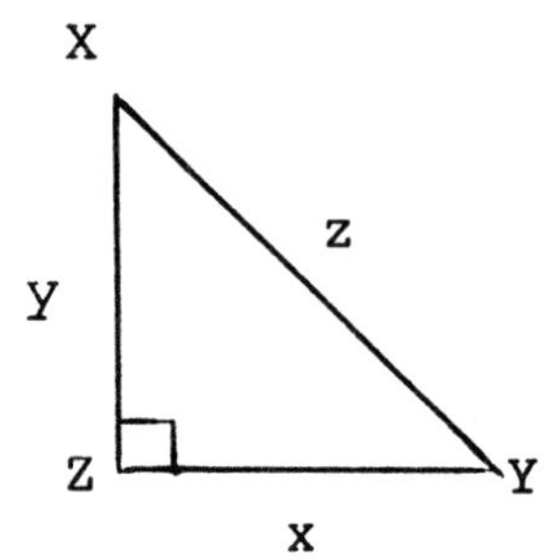

Analysis: First identify the hypothesis A and the conclusion B, so

A: The right triangle XYZ has an area of $z^2/4$.

B: Triangle XYZ is isosceles.

<u>The Backward Process</u>

Looking at statement B, you should ask the

Key Question: How can I show that a triangle is isosceles?

One possible answer is the

Key Answer: Show that two sides of the triangle have equal length.

Applying the key answer to the specific problem, you obtain the new statement to be proved:

B1: $x = y$.

All efforts should now be directed toward reaching the conclusion that B1 is true. To do so, you must eventually work forward from the hypothesis A; however, to illustrate some of the difficulties that can arise during the backward process, let's continue working backward from B1. Can you pose a key question for B1? Certainly, one could ask the

Key Question: How can I show that the lengths of two sides of a triangle are equal?

Another perfectly valid key question would be the

Key Question: How can I show that two real numbers are equal?

After all, x and y are real numbers. When more than one key question is available, try to use the information in the hypothesis to choose the correct one. For instance, if you selected the first key question above, then an associated key answer would be the

Key Answer: Show that the angles opposite the two sides are equal.

Now, the hypothesis will be used to guide you in choosing the key question. Since the hypothesis A does not seem to contain much information relating to the angles of triangle XYZ, the second key question should be chosen.

Another difficulty with the backward process is that a key question can have many key answers. For instance, associated with the second key question above could be either of the following:

Key Answer: Show that the difference of the two numbers is 0.

Key Answer: Show that the first number is less than or equal to the second number, and that the second number is less than or equal to the the first one.

For the sake of argument, suppose that the first key answer is chosen. Applying that answer to the specific problem, you obtain the new statement to be proved:

B2: $x - y = 0$.

Observe that if B2 is true, then so is B1, and hence B. All efforts are now directed toward establishing that B2 is true. Continuing the backward process, from B2 you could ask the

Key Question: How can I show that the difference of two real numbers is 0?

It might seem that this key question does not have an obvious answer. When that happens, or when you can no longer pose and/or answer the key question fruitfully, turn to the forward process.

The Forward Process

The objective now is to work forward from the hypothesis A to obtain statement B2. Looking at A, and using the fact that the area of a triangle is one-half the base times the height, you can state

A1: $xy/2 = z^2/4$.

Another useful statement follows from A by the Pythagorean theorem, so you also have

A2: $x^2 + y^2 = z^2$.

It is possible to combine A1 and A2 by replacing z^2 in A1 with $x^2 + y^2$ from A2 obtaining

A3: $xy/2 = (x^2 + y^2)/4$.

Recall that the objective of the forward process is to obtain statement B2: $x - y = 0$. Note that this statement does not contain z. It was for this reason that z^2 was eliminated from A1 and A2.

The remainder of the forward process is to apply algebraic manipulations to A3 so as to obtain precisely B2. Specifically, you can multiply both sides of A3 by 4 and subtract 2xy from both sides to obtain

A4: $x^2 - 2xy + y^2 = 0$.

By factoring you can obtain

A5: $(x - y)^2 = 0$.

The final step of the forward process, and in the entire proof, is to realize, from A5, that the only way for the number x - y times itself to be 0 is if x - y = 0, thus obtaining precisely

B2: $x - y = 0$.

Note that A5 ultimately produced the answer to the key question associated with B2: "How can I show that the difference of two numbers is 0?" Looking at A5, the answer is to show that the square of the difference of the two numbers is 0.

Summary of this Proof

The steps of this proof will now be summarized. Observe the complete lack of reference to the backward process and to the key question.

PROOF

STATEMENT	REASON
1. Area of the right triangle XYZ is $z^2/4$.	1. Given in the hypothesis.
2. $xy/2 = z^2/4$.	2. Area = (base)(height)/2.
3. $x^2 + y^2 = z^2$.	3. Pythagorean theorem.
4. $xy/2 = (x^2 + y^2)/4$.	4. Substitute (3) into (2).
5. $x^2 - 2xy + y^2 = 0$.	5. Algebra.
6. $(x - y)^2 = 0$.	6. Factoring (5).
7. $(x - y) = 0$.	7. From (6).
8. $x = y$.	8. Add y to both sides of (7).
9. Triangle XYZ is isosceles.	9. Definition of isosceles triangle.

EXERCISES FOR SECTION 2.5

For each of the following conclusions, list two different key questions that contain no symbols from the specific problem.

1. If R and S are two tangent lines to a circle C at the two endpoints E and F of a diameter D, respectively, then R and S are parallel.

   Key Question: How can I show that ____________________

   ____?

   Key Question: How can I show that ____________________

   ____?

2. If f(x) and g(x) are polynomials, then the function f(x) + g(x) is a polynomial.

   Key Question: How can I show that ____________________

   ____?

   Key Question: How can I show that ____________________

   ____?

For each of the following key questions, list three different key answers.

3. Key Question: How can I show that two triangles are congruent?

Key Answer: Show that ______________________________

Key Answer: Show that ______________________________

Key Answer: Show that ______________________________

4. Key Question: How can I show that two lines are parallel?

Key Answer: Show that ______________________________

Key Answer: Show that ______________________________

Key Answer: Show that ______________________________

5. Key Question: How can I show that a quadrilateral is a rectangle?

Key Answer: Show that ______________________________

Key Answer: Show that ______________________________

Key Answer: Show that ______________________________

For each of the following conclusions, (1) pose a key question, (2) provide a key answer, and (3) apply your answer to the specific problem to obtain the new statement B1 that is to be proved.

6. If a, b, and c are real numbers for which $a > 0$, $b < 0$, and $b^2 - 4ac = 0$, then the solution to the equation $ax^2 + bx + c = 0$ is positive.

Key Question: How can I show that ______________________

____?

Key Answer: Show that ______________________

____

B1: ______________________________________

7. If SU is a perpendicular bisector of RT, and RS = 2RU, then triangle RST is equilateral.

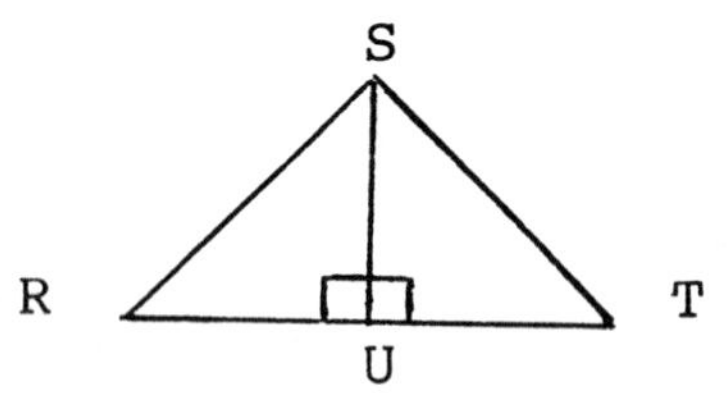

Key Question: How can I show that ______________________

____?

Key Answer: Show that ______________________

____

B1: ______________________________________

For each of the following statements A, list three different statements that could be obtained as a result of applying the forward process for precisely one step from A.

8. A: The circle C consists of all values for x and y that satisfy $(x - 3)^2 + (y - 2)^2 = 25$.

A1: __________ A1: __________ A1: __________

*9. A: The sine of $\angle$ X in triangle XYZ of Example 3 is $1/\sqrt{2}$.

A1: __________ A1: __________ A1: __________

10. A: The triangle UVW is equilateral.

A1: __________ A1: __________ A1: __________

11. Circle the statement that cannot be obtained by working forward from the hypothesis of the problem, "If x and y are real numbers such that $x^2 + 6y^2 - 25 = 0$ and $y^2 + x = 3$, then $|y| = 2$."

(a) $y^2 = 3 - x$

(b) $y^2 = 25/6 - (x/\sqrt{6})^2$

(c) $(3 - y^2)^2 + 6y^2 - 25 = 0$

(d) $(x + 5) = -6y^2/(x - 5)$

Explain your choice. ______________________________

____

2.6 HOW TO WRITE A PROOF

After solving a problem, it is necessary to write down your proof so that it can be communicated to someone else. The form of the presentation that is usually used in geometry does not include all of the thought process that went into solving the problem, because that would require too much time, effort, and space. Rather, a tabular summary is provided, as was done at the end of Examples 1, 2, and 3.

A formal presentation of a proof in geometry should consist of:

1. Problem: a full written statement of the problem.

2. Figure: a figure with symbols and notation to illustrate all the given conditions.

3. Given: a statement, using the notation of the figure, indicating what information is given in the hypothesis.

4. To Prove: a statement, using the notation of the figure, indicating the desired conclusion.

5. Analysis: a detailed description of which proof techniques are being used in the proof, and how they are being applied. (The forward-backward method is one available technique. Other proof techniques will be described in later chapters.)

6. Summary Table: a list of numbered statements and correspondingly numbered reasons that lead from the hypothesis to the conclusion. In general, the summary table contains no reference to the backward process or to the key questions.

For now, your proofs should include all six of these components. In the exercises of each section, your proofs should contain an analysis that is similar to the ones in the text, although perhaps a bit shorter. Later in your mathematics career, the analysis and summary table will be replaced with a proof that is written in paragraph form. Such a proof is generally not as detailed as the summary table. Several steps are often combined into one sentence. For this reason, a written version of a proof will be called a <u>condensed</u> proof. There are no standard rules for writing a condensed proof. The fewer details that are given, the harder it will be for someone else to read and to understand the condensed proof. Here are some condensed proofs for Examples 1, 2, and 3.

Condensed Proof of Example 1: From the hypothesis, $\angle$ RST and $\angle$ RSU are right angles. Thus, by the definition of right angles, $\angle RST = \angle RSU = 90^\circ$. On adding the degrees of these two angles, one has $\angle RST + \angle RSU = 180^\circ$. From the definition of supplementary angles, it follows that $\angle$ RST and $\angle$ RSU are supplementary.//

Note: The "//" or some equivalent symbol is used to indicate the end of the proof. Sometimes, the letters Q.E.D. are used as they stand for the Latin words <u>quod erat demonstrandum</u>, meaning, "which was to be demonstrated."

Condensed Proof of Example 2: From the hypothesis that RT is perpendicular to ST, and by the definition of perpendicularity, $\angle 3 = 90^\circ$. Also, it is given in the hypothesis that $\angle 1 + \angle 2 + \angle 3 = 180^\circ$. By subtracting, one obtains $\angle 1 + \angle 2 = 90^\circ$. Consequently, by definition, it follows that $\angle 1$ and $\angle 2$ are complementary.//

*Condensed Proof of Example 3: From the hypothesis and the formula for the area of a right triangle, the area of XYZ = $xy/2 = z^2/4$. By the Pythagorean theorem, $x^2 + y^2 = z^2$ and, on substituting $(x^2 + y^2)$ for z^2 and performing some algebraic manipulations, one obtains $(x - y)^2 = 0$. Hence $x = y$ and the triangle XYZ is isosceles.//

Note that these proofs have been written completely from the forward process. Other ways of writing condensed proofs will be presented in the next section.

EXERCISES FOR SECTION 2.6

Rewrite the following proofs in condensed form.

1. Problem: If the line RS intersects the lines TU and VW in such a way that $\angle 3 = \angle 4$, then $\angle 1$ and $\angle 2$ are supplementary.

Figure:

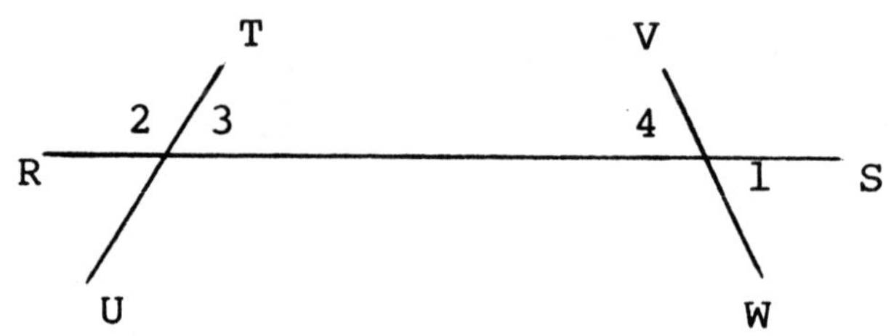

Given: The line RS intersects the lines TU and VW, and $\angle 3 = \angle 4$.

To Prove: $\angle 1$ and $\angle 2$ are supplementary.

Analysis: The forward-backward method will be used together with the properties of supplementary angles.

Summary Table:

PROOF

STATEMENT	REASON
1. RS is a line.	1. Given.
2. $\angle 2 + \angle 3 = 180^\circ$.	2. Because $\angle 2 + \angle 3 = \angle RS$, and RS is a line, so $\angle RS = 180^\circ$.
3. $\angle 3 = \angle 4$.	3. Given.
4. $\angle 2 + \angle 4 = 180^\circ$.	4. Substituting (3) into (2).
5. $\angle 4 = \angle 1$.	5. Vertical angles of intersecting lines are equal. (Recall that $\angle 4$ and $\angle 1$ are referred to as vertical angles.)
6. $\angle 2 + \angle 1 = 180^\circ$.	6. Substituting (5) into (4).
7. $\angle 2$ and $\angle 1$ are supplementary.	7. Definition of supplementary angles.

Condensed Proof: ______________________________

2. Problem: If two angles ($\angle$ 1 and $\angle$ 3) are complementary to equal angles ($\angle$ 2 and $\angle$ 4), then they are equal.

Figure:

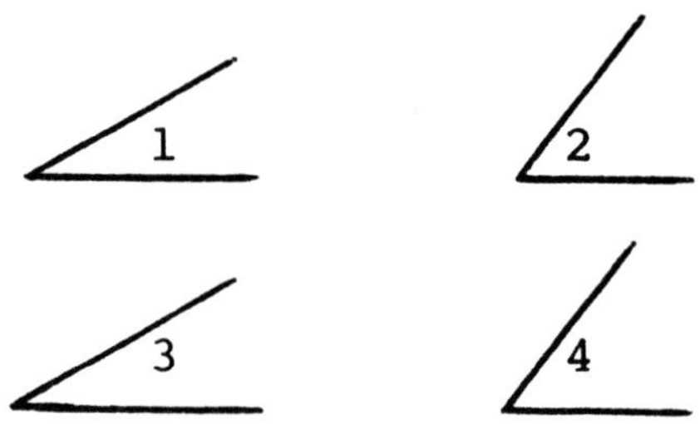

Given: $\angle$ 1 and $\angle$ 2 are complementary.

$\angle$ 3 and $\angle$ 4 are complementary.

$\angle 2 = \angle 4$.

To Prove: $\angle 1 = \angle 3$.

Analysis: The forward-backward method will be used together with the properties of complementary angles.

Summary Table:

PROOF

STATEMENT	REASON
1. $\angle 1$ and $\angle 2$ are complementary.	1. Given.
2. $\angle 1 + \angle 2 = 90^\circ$.	2. Definition of complementary angles.
3. $\angle 3$ and $\angle 4$ are complementary.	3. Given.
4. $\angle 3 + \angle 4 = 90^\circ$.	4. Definition of complementary angles.
5. $\angle 1 + \angle 2 = \angle 3 + \angle 4$.	5. Both numbers are equal to 90°.
6. $\angle 2 = \angle 4$.	6. Given.
7. $\angle 1 = \angle 3$.	7. Subtracting (6) from (5).

Condensed Proof: ____

____

____

____

____

____

____

____

For each of the following problems, supply all of the missing information. When necessary, use the forward-backward method to solve the problem.

3. Problem: If $\angle$ TSU = $\angle$ USV and RU is a line, then $\angle$ USV and $\angle$ RST are supplementary.

Figure:

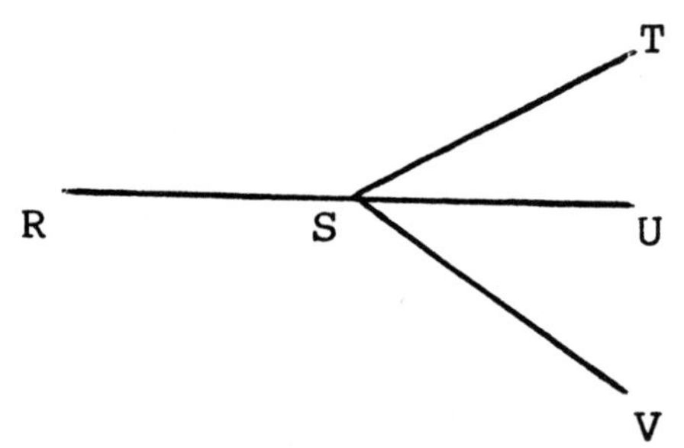

Given: ____

To Prove: ____

Analysis: The forward-backward method will be used together with the property of supplementary angles.

Summary Table:

PROOF

STATEMENT	REASON
1. RU is a line.	1. Given.
2. $\angle$ RST + $\angle$ TSU = 180°.	2. ______
3. $\angle$ TSU = $\angle$ USV.	3. ______
4. $\angle$ RST + $\angle$ USV = 180°.	4. ______
5. $\angle$ RST and $\angle$ USV are supplementary.	5. ______

Condensed Proof: ______

4. Problem: If RS $\perp$ ST, and ST $\perp$ TU, and $\angle$ 1 = $\angle$ 2, then $\angle$ 3 = $\angle$ 4.

Figure:

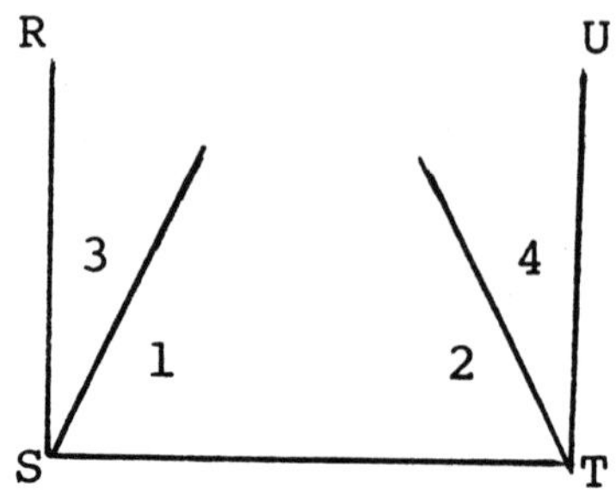

Given: ____________________

____

____

To Prove: ____________________

Analysis: The forward-backward method will be used together with facts about perpendicularity.

Summary Table: (Note: Use as many lines as you need.)

PROOF

STATEMENT	REASON
1. __________	1. __________
2. __________	2. __________
3. __________	3. __________
4. __________	4. __________
5. __________	5. __________
6. __________	6. __________
7. __________	7. __________
8. __________	8. __________
9. __________	9. __________
10. __________	10. __________

5. Problem: If two lines intersect, then the vertical angles formed are equal.

Figure:

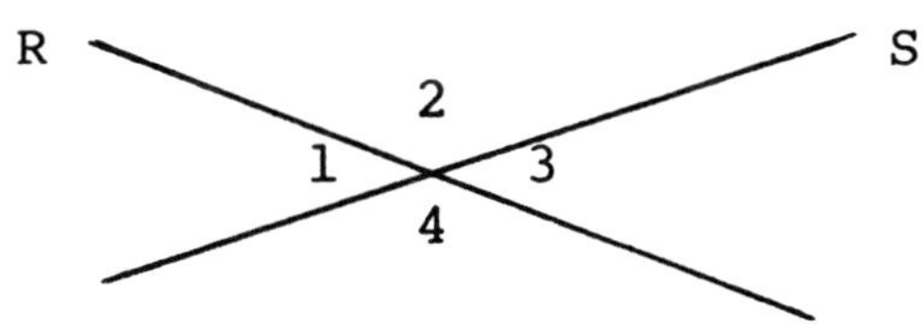

Given: ____________________

To Prove: ____________________

Analysis: The forward-backward method will be used together with facts about supplementary angles.

Summary Table: (Note: Use as many lines as you need.)

PROOF

STATEMENT	REASON
1. ______	1. ______
2. ______	2. ______
3. ______	3. ______
4. ______	4. ______
5. ______	5. ______
6. ______	6. ______
7. ______	7. ______
8. ______	8. ______
9. ______	9. ______
10. ______	10. ______

Condensed Proof: ______

______

______

______

______

______

______

______

6. Problem: If $RS \perp RU$, and $ST \perp TU$, and $\angle 1 = \angle 2$, then $\angle 3 = \angle 4$.

Forward-Backward

Figure:

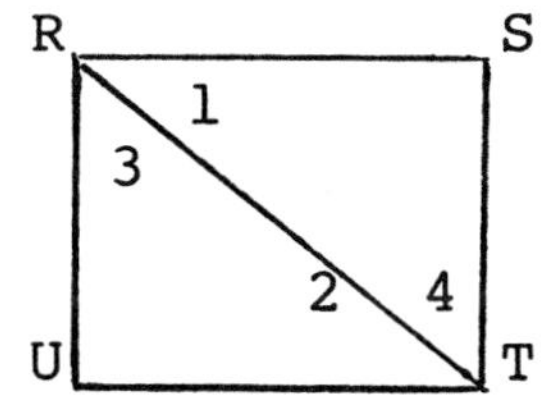

Given: ________________________

____

____

To Prove: ________________________

Analysis: Use the forward-backward method together with facts about perpendicularity.

Summary Table: (Note: Use as many lines as you need.)

PROOF

STATEMENT	REASON
1.	1.
2.	2.
3.	3.
4.	4.
5.	5.
6.	6.
7.	7.
8.	8.
9.	9.
10.	10.

Condensed Proof: ______________________________

____

____

____

____

____

____

____

*7. Problem: If RU is a perpendicular bisector of ST, and RS = 2SU, then triangle RST is equilateral.

Figure:

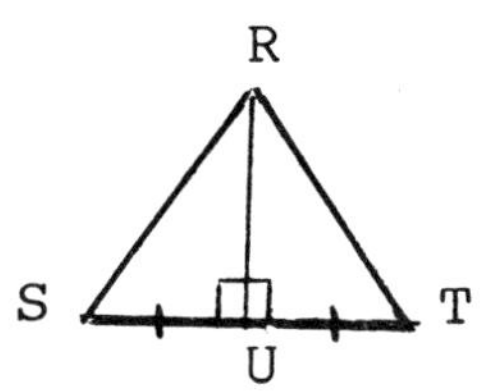

(Note: Supply a very detailed analysis of your proof. Indicate the key questions and answers.)

Analysis: ______________________________

____

____

____

____

____

____

Condensed Proof:

*8. Problem: If the right triangle XYZ is isosceles, then the area of the triangle is $z^2/4$.

Figure:

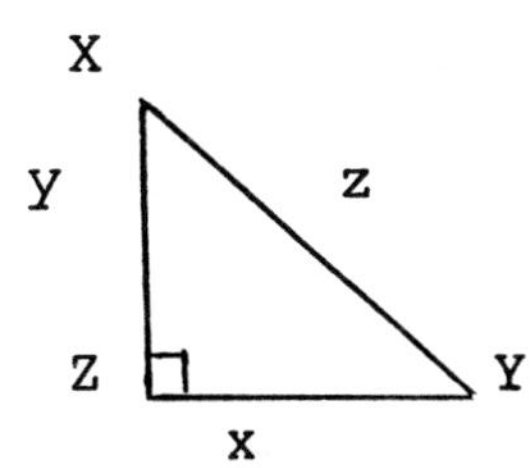

(Note: Supply a very detailed analysis of your proof. Indicate the key questions and answers.)

Analysis:

Condensed Proof:

2.7 HOW TO READ A PROOF

When a proof is presented in a table summary, it is usually possible to understand the reasoning of each step. However, when a proof is presented in condensed form, it can be more difficult to follow the author's reasoning because

1. Most condensed proofs make little or no reference to the different proof techniques that are being used, particularly to the backward process and to the key questions.

2. Two, three, or more steps of the proof are often condensed into a single sentence and, sometimes, little or no justification is given to support these statements.

In order to read a condensed proof, you must learn to

1. Identify which techniques are being used, because the forward-backward method is not the only one available.

2. Fill in the necessary steps and reasons when the author has omitted them.

Note: If you encounter difficulties in reading a proof, proceed by asking yourself how you would do the proof, and then see if the author is following the same reasoning.

Here are some condensed proofs of Examples 1, 2, and 3.

Condensed Proof of Example 1: To reach the desired conclusion, it will be shown that $\angle RST + \angle RSU = 180^\circ$. From the hypothesis, $\angle RST$ and $\angle RSU$ are right angles. It then follows that $\angle RST + \angle RSU = 90^\circ + 90^\circ = 180^\circ$.//

Discussion: The first sentence of the proof is written from the backward process while, in the second sentence, the author has turned to the forward process. In the third sentence several steps have been skipped when the author says, "It follows that ...."

Condensed Proof of Example 2: To establish that the two angles are complementary, it will be shown that their degrees add up to 90°. Since RT is perpendicular to ST, $\angle 3 = 90^\circ$. Also, by hypothesis, $\angle 1 + \angle 2 + \angle 3 = 180^\circ$. Hence, $\angle 1 + \angle 2 = 90^\circ$.//

Discussion: Again, the first sentence indicates that the backward process is being discussed, while the second and third sentences are written from the forward process. Finally, the word "hence" in the fourth sentence suggests that some steps and reasons have been omitted.

*Condensed Proof of Example 3: (Partly backward, partly forward). The statement will be proved by establishing that $x = y$, which in turn is done by showing that $(x - y)^2 = (x^2 - 2xy + y^2) = 0$. But the area of the triangle is $(1/2)xy = (1/4)z^2$, so that $2xy = z^2$. By the Pythagorean theorem, $z^2 = (x^2 + y^2)$, and hence $(x^2 + y^2) = 2xy$, or $(x^2 - 2xy + y^2) = 0$, as required.//

*Condensed Proof of Example 3: (All backward). To reach the conclusion, it will be shown that $x = y$ by verifying that $(x - y)^2 = (x^2 - 2xy + y^2) = 0$ or, equivalently, that $(x^2 + y^2) = 2xy$. This can be established by showing that $2xy = z^2$, for the Pythagorean theorem states that $(x^2 + y^2) = z^2$. In order to see that $2xy = z^2$ or, equivalently, that $(1/2)xy = (1/4)z^2$, note that $(1/2)xy$ is the area of the triangle and it is equal to $(1/4)z^2$ by hypothesis, thus completing the proof.//

*Condensed Proof of Example 3: (Very condensed). The hypothesis together with the Pythagorean theorem yield $x^2 + y^2 = 2xy$, hence $(x - y)^2 = 0$. Thus the triangle is isosceles as required.//

EXERCISES FOR SECTION 2.7

For each sentence in the following condensed proofs, indicate whether the sentence relates to the forward or backward process by circling the appropriate word.

1. Problem: If $\angle 1 + \angle 2 + \angle 3 = 180^\circ$, and $\angle 1$ are $\angle 2$ are complementary, then triangle XYZ is a right triangle.

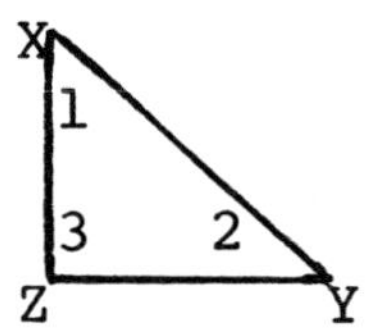

Condensed Proof:

S1: Since $\angle 1$ and $\angle 2$ are complementary, $\angle 1 + \angle 2 = 90^\circ$. (Forward or Backward)

S2: To see that triangle XYZ is a right triangle, it will be shown that $\angle 3 = 90^\circ$. (Forward or Backward)

S3: But since $\angle 1 + \angle 2 + \angle 3 = 180^\circ$ by the hypothesis, and $\angle 1 + \angle 2 = 90^\circ$, by subtracting it follows that $\angle 3 = 90^\circ$.// (Forward or Backward)

2. Problem: If lines R and S are perpendicular, then $\angle 1$ and $\angle 2$ are supplementary.

Forward-Backward

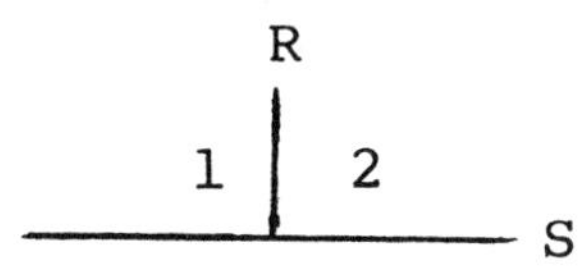

Condensed Proof:

S1: To reach the desired conclusion, it will be shown that $\angle 1 + \angle 2 = 180^\circ$. (Forward or Backward)

S2: This in turn will be accomplished by showing that both $\angle 1$ and $\angle 2$ equal 90°, for then, $\angle 1 + \angle 2 = 90^\circ + 90^\circ = 180^\circ$. (Forward or Backward)

S3: By the hypothesis that $R \perp S$, $\angle 1$ and $\angle 2$ are right angles. (Forward or Backward)

S4: But by the definition of right angles, this means that $\angle 1 = 90^\circ$ and $\angle 2 = 90^\circ$, thus completing the proof.// (Forward or Backward)

3. Problem: If $\angle$ TSU = $\angle$ USV and RU is a line, then $\angle$ USV and $\angle$ RST are supplementary.

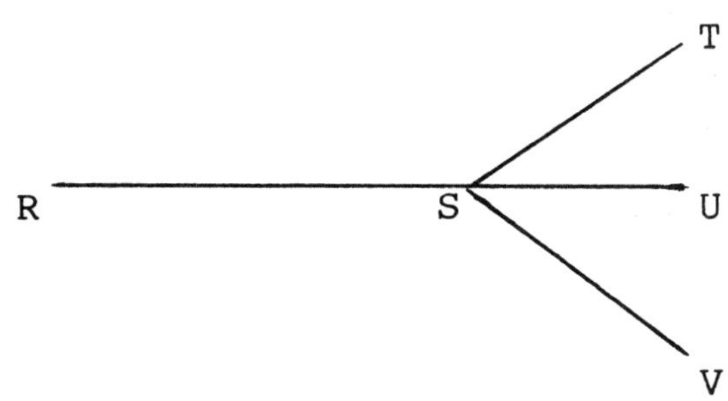

Forward-Backward

Condensed Proof:

S1: Since RU is a line and $\angle$ RST + $\angle$ TSU = $\angle$ RU, it follows that $\angle$ RST + $\angle$ TSU = 180°. (Forward or Backward)

S2: To see that $\angle$ USV and $\angle$ RST are supplementary, it will be shown that $\angle$ USV + $\angle$ RST = 180°. (Forward or Backward)

S3: But since $\angle$ TSU = $\angle$ USV, and since it has already been established that $\angle$ RST + $\angle$ TSU = 180°, it follows that $\angle$ RST + $\angle$ USV = 180°, thus completing the proof.// (Forward or Backward)

For the following condensed proofs, provide the associated summary table. When necessary, add steps to fill in the missing details.

4. Problem: If $\angle$ 1 and $\angle$ 2 are complementary, and $\angle$ 3 and $\angle$ 4 are complementary, and $\angle$ 2 = $\angle$ 4, then $\angle$ 1 = $\angle$ 3.

Condensed Proof: By the hypothesis, $\angle 1$ and $\angle 2$ are complementary, so $\angle 1 + \angle 2 = 90^\circ$. Similarly, since $\angle 3$ and $\angle 4$ are complementary, $\angle 3 + \angle 4 = 90^\circ$. Thus, $\angle 1 + \angle 2 = \angle 3 + \angle 4$. Since the hypothesis states that $\angle 2 = \angle 4$, by subtracting, one reaches the conclusion that $\angle 1 = \angle 3$.//

Summary Table:

PROOF

	STATEMENT		REASON
1.		1.	
2.		2.	
3.		3.	
4.		4.	
5.		5.	
6.		6.	
7.		7.	
8.		8.	
9.		9.	
10.		10.	

5. Problem: If RS if perpendicular to ST, and $\angle 1 = \angle 3$ then XS is perpendicular to SY.

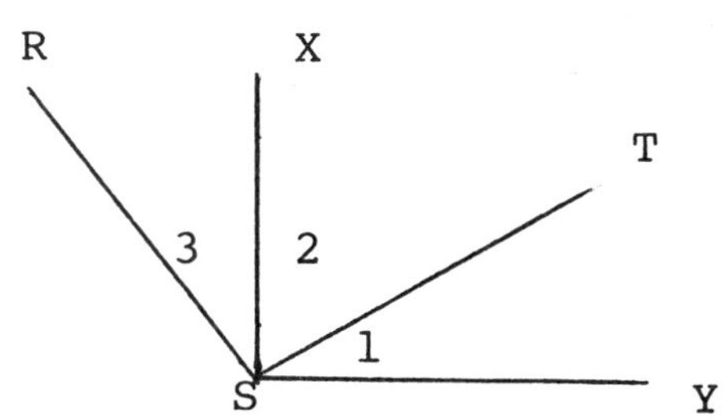

Forward-Backward

Condensed Proof: To reach the desired conclusion, it will be shown that $\angle$ XSY = 90°, for then, XS will be perpendicular to SY. By the hypothesis that RS is perpendicular to ST, $\angle$ RST = 90°, and then $\angle$ 3 + $\angle$ 2 = $\angle$ RST = 90°. Finally, the fact that $\angle$ 3 = $\angle$ 1 can be used to claim that $\angle$ 1 + $\angle$ 2 = 90°, as desired.//

Summary Table:

PROOF

	STATEMENT		REASON
1.		1.	
2.		2.	
3.		3.	
4.		4.	
5.		5.	
6.		6.	
7.		7.	
8.		8.	
9.		9.	
10.		10.	

6. Problem: If RS and TU are two intersecting lines, then $\angle$ 1 = $\angle$ 3.

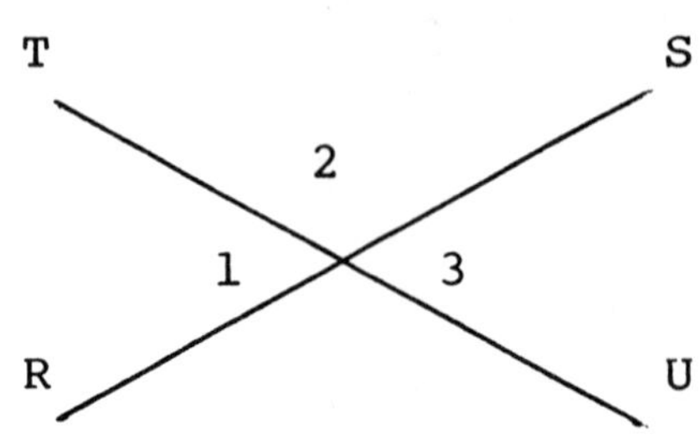

Condensed Proof: Since RS is a line, $\angle 1 + \angle 2 = \angle RS = 180^\circ$. Similarly, since TU is a line, $\angle 2 + \angle 3 = \angle TU = 180^\circ$. Consequently, $\angle 1 = \angle 3$, as desired.//

Summary Table:

PROOF

STATEMENT	REASON
1. ______	1. ______
2. ______	2. ______
3. ______	3. ______
4. ______	4. ______
5. ______	5. ______
6. ______	6. ______
7. ______	7. ______
8. ______	8. ______
9. ______	9. ______
10. ______	10. ______

7. Problem: If RS is a line, and $\angle 2 = \angle 3$, then $\angle 1$ and $\angle 4$ are supplementary.

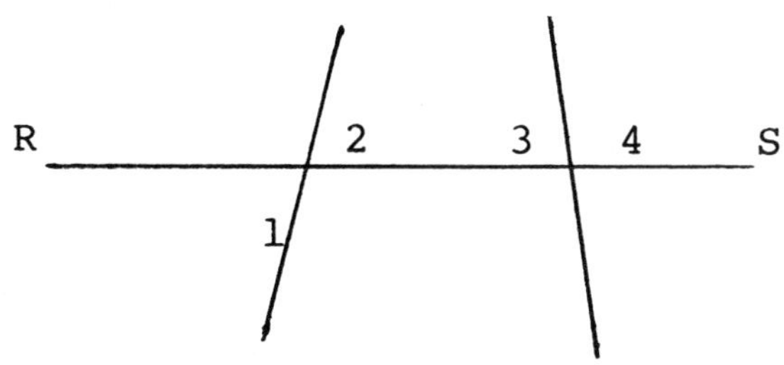

Condensed Proof: It will be shown that $\angle 1 + \angle 4 = 180^\circ$ by establishing that $\angle 3 + \angle 4 = 180^\circ$ and $\angle 3 = \angle 1$. To see that $\angle 3 + \angle 4 = 180^\circ$, note that RS is a line, and so $\angle 3 + \angle 4 = \angle RS = 180^\circ$. Finally, to see that $\angle 3 = \angle 1$, note that $\angle 1 = \angle 2$ since vertical angles of intersecting lines are equal, and by the hypothesis, $\angle 2 = \angle 3$. Thus the proof is complete.//

Summary Table:

PROOF

STATEMENT	REASON
1. ____	1. ____
2. ____	2. ____
3. ____	3. ____
4. ____	4. ____
5. ____	5. ____
6. ____	6. ____
7. ____	7. ____
8. ____	8. ____
9. ____	9. ____
10. ____	10. ____

2.8 SUMMARY

The forward-backward method is one of several proof techniques for showing that,

If A is true, then B is true.

Since it is usually best to start a proof with the backward process, it will be summarized before the forward process.

Summary of the Backward Process

Through asking and answering the key question, derive from B a new statement, B1, with the property that if B1 is true, then so is B. Continue the backward process from B1 until either you reach statement A, or until you can no longer pose and/or answer the key question fruitfully. Then turn to the forward process.

Summary of the Forward Process

Starting from the hypothesis A, derive other statements, A1, A2, ..., that are true as a result of A being true. Continue this process until you obtain precisely the last statement of the backward process, at which time the proof will be complete. Like a column of dominoes, you can do the proof by going from A all the way to B.

Summary of How to Write a Proof

After carefully drawing a figure with symbols and notation to represent the problem, identify the hypothesis (Given) and the conclusion (To Prove). Provide a detailed analysis indicating which proof techniques will be used, such as the forward-backward method, and how they are being applied. Finally, provide a summary table of the steps and the associated reasons. Eventually you must learn to write a condensed proof in paragraph form.

Summary of How to Read a Proof

When reading a condensed proof, learn to identify which techniques are being used, even if the author does not clearly state them. When several steps are omitted, try to recreate the author's thoughts and to provide the necessary reasons for each step. If you are having difficulties, then ask yourself how you would do the proof. Then see if the author is following the same reasoning.

CHAPTER 3
On Definitions and Mathematical Terminology

3.1 DEFINITIONS AND BASIC TERMINOLOGY

In the previous chapter, you learned the forward-backward method and you saw the importance of formulating and answering the key question. One of the simplest yet most effective ways of answering a key question is through the use of a definition, as you will soon learn.

Undefined Terms

Some terms in geometry such as "point," "line," and "plane," are left undefined although their meaning is clear. In set theory, a set can be described as a collection of objects. Once again, the term "objects" is left undefined. Any attempt to define these terms would result in the use of other terms that, in turn, would have to be defined, and so on, and so on. Eventually, one would have to accept certain terms without giving them a formal definition. These undefined terms then form the building blocks for defining other mathematical terms.

Definitions

A definition is a statement that is agreed on by all parties concerned. Definitions are not made randomly. Usually, they are motivated by a mathematical concept that occurs repeatedly. For example:

Definition: A triangle is <u>equilateral</u> if all three of its sides have equal length.

Definitions and Terminology

Surely it is easier to say "an equilateral triangle" than "a triangle in which all three of its sides have equal length," especially if the concept arises frequently. In fact, a definition can be viewed as an abbreviation for a particular mathematical concept that is agreed on by all parties concerned. A good definition should use only accepted undefined terms or previously defined terms. Some other examples of definitions would be:

Definition 1: Two angles are supplementary if their degrees add up to 180°.

Definition 2: Two angles are complementary if their degrees add up to 90°.

Definition 3: An angle is a right angle if it has 90°.

Definition 4: Two intersecting lines are perpendicular if the angles formed at their intersection are all right angles.

Definition 5: A triangle is isosceles if at least two of its sides have equal length.

Definition 6: A triangle is a right triangle if one of its angles is a right angle.

Definition 7: An integer n is even if n = 2k for some integer k.

Definition 8: An integer n is odd if n = 2k + 1 for some integer k.

Definition 9: A real number r is a rational number if r can be expressed as the ratio of two integers p and q in which the denominator q is not 0.

Definition 10: Two lines are <u>parallel</u> if they lie in the same plane and they do not intersect.

From the examples above, you can see that an acceptable definition must

1. Name the term being defined, but never use the term within the actual definition.

2. Use only previously defined terms or accepted undefined terms.

3. State those (and only those) characteristics that distinguish the defined term from other objects.

4. Contain the least possible amount of information.

5. Be reversible -- for example, not only is it the case that an equilateral triangle is a triangle with all three sides of equal length, but also, if a triangle has all three sides of equal length, then the triangle is equilateral. More will be said about the issue of reversibility in Section 3.3.

The following definition of a square is not acceptable because it does not state enough characteristics to distinguish a square from other geometric figures that are not squares.

Definition: A <u>square</u> is a four-sided figure in a plane that has all four sides of equal length.

The following figure satisfies the above definition but it is not a square.

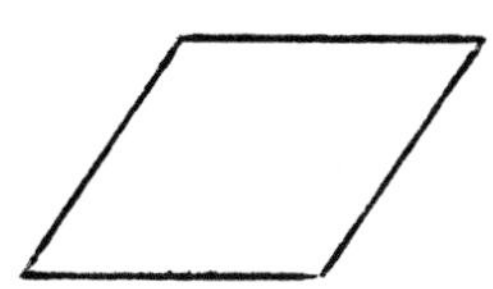

Definitions and Terminology

Postulates

Just as there are certain terms that are accepted as undefined, so are there certain statements that are accepted as being true without a proof. These unproved statements are referred to as postulates, or axioms. Some examples of postulates would be:

Postulate P1: There is exactly one line through two different points.

Postulate P2: Given a line and a point not on the line, there is one and only one line through the point and parallel to the given line.

A postulate can be used in the forward and backward processes to support a step in a proof.

Basic Terminology

In dealing with proofs, there are four terms that you will often come across. They are:

1. Proposition -- a true statement of interest that you are trying to prove.

2. Theorem -- a proposition that is considered (subjectively) to be very important.

3. Lemma -- a preliminary proposition that is to be used in the proof of a theorem.

4. Corollary -- a proposition that follows almost immediately as the result of knowing that a theorem is true.

EXERCISES FOR SECTION 3.1

Indicate whether the following definitions are acceptable or not. If not, explain which of the five properties of an acceptable definition are not being satisfied.

Example: A square is a four-sided figure in a plane that has four equal sides, four equal angles and whose diagonals are perpendicular.

Answer: Unacceptable -- does not contain the least amount of information.

1. A circle is a round, circular object.

   ____

   ____

2. The bisector of an angle is a ray that divides an angle into two angles having the same number of degrees.

   ____

   ____

3. A line segment is a set of points that lie evenly between two points.

   ____

   ____

4. A <u>median</u> of a triangle is a line segment joining a vertex to the opposite side.

____

____

5. An <u>obtuse angle</u> is an angle that has more than 90° and is not acute.

____

____

Rewrite Definitions 7, 8, and 9 so that there are no symbols in the statement.

6. Definition 7: An integer is <u>even</u> if ________________

____

____

7. Definition 8: An integer is <u>odd</u> if ________________

____

____

8. Definition 9: A real number is a <u>rational number</u> if

____

____

Definitions and Terminology

For each of the following groups of pictures, look for a common property and use that property to define the associated term. You may use accepted undefined terms and any term defined in this section, as well as the concept of a polygon.

Example:

Acute Angles

Not Acute Angles

Answer: An acute angle is an angle with less than 90°.

9. Quadrilaterals

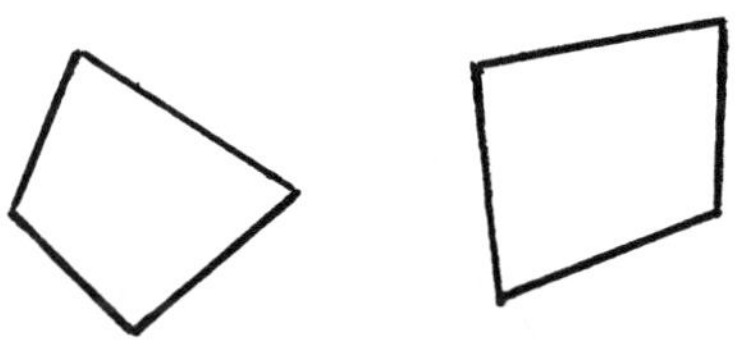

Not Quadrilaterals

A quadrilateral is ______________________________

____

____

10. Parallelograms — Not Parallelograms

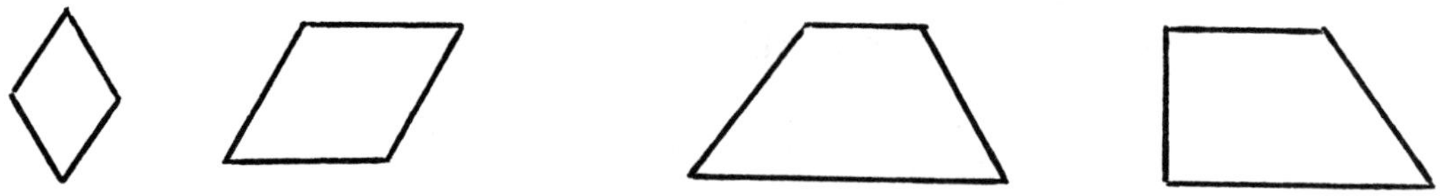

A parallelogram is ______________________________

____

____

11. Rectangles — Not Rectangles

A rectangle is ______________________________

____

____

*12. Convex Polygons — Not Convex Polygons

A convex polygon is ______________________________

____

____

3.2 USING DEFINITIONS IN PROOFS

In the proof of Example 1, a definition was used to answer a key question. Recall the very first

Key Question: How can I show that two angles are supplementary?

The answer was obtained from Definition 1, which led to the

Key Answer: Show that the degrees of the two angles add up to 180°.

Definitions are equally useful in the forward process. For instance, if you know that an integer n is odd then, by Definition 8, you would know that $n = 2k + 1$ for some integer k.

Warning: When applying a definition that contains symbols to a specific problem, be sure to apply the definition carefully using the notation of the specific problem.

Example: If you know that an integer m^2 is odd then, by Definition 8, you could work forward and say that $m^2 = 2p + 1$ for some integer p. (Note that p has been used instead of k, which is perfectly acceptable.)

The next example shows the use of definitions in both the forward and backward processes.

Example 4: If n is an even integer, then n^2 is an even integer.

Given: n is an even integer.

To Prove: n^2 is an even integer.

Analysis: The forward-backward method will be used to do the proof, together with the definition of an even integer.

The Backward Process

The backward process gives rise to the

Key Question: How can I show that an integer (namely n^2) is an even integer?

Using Definition 7 leads to the

Key Answer: Show that the integer equals two times some other integer.

Applying the key answer to the specific integer n^2, the result of the backward process is the new statement to be proved:

B1: $n^2 = 2k$ for some integer k.

In order to show that B1 is true, you must determine the specific integer k for which $n^2 = 2k$. To do so, turn to the forward process.

The Forward Process

The hypothesis of this problem states that n is an even integer. By using Definition 7, one step in the forward process results in the statement

A1: $n = 2p$ for some integer p.

Note that the symbol p was used instead of k, because k was already used in the backward process. Using different symbols in this way can often avoid confusion.

The objective now is to use the integer p from A1 to obtain the value for the integer k that is needed in B1. Specifically, since B1 deals with n^2 and A1 deals with n, why not square both sides of the equality in A1 to obtain

A2: $n^2 = (2p)^2 = (2p)(2p) = 4p^2$.

Finally, A2 can be rewritten as

A3: $n^2 = 2(2p^2)$.

Looking carefully at A3 and at B1, it should not be difficult to see that the desired value for the integer k is $2p^2$, thus completing the proof.

Summary Table:

PROOF

STATEMENT	REASON
1. n is an even integer.	1. Given.
2. $n = 2p$ for some integer p.	2. Definition of an even integer.
3. $n^2 = (2p)^2$.	3. Square both sides of (2).
4. $n^2 = 4p^2 = 2(2p)^2$.	4. Algebra applied to (3).
5. $n^2 = 2k$, where $k = 2p^2$.	5. Rewriting (4).
6. n^2 is even.	6. Definition of an even integer.

Condensed Proof: Since the hypothesis states that n is an even integer, by definition, it follows that $n = 2p$ for some integer p. But then

$$n^2 = (2p)^2 = 4p^2 = 2(2p^2),$$

and so n^2 is an even integer.//

EXERCISES FOR SECTION 3.2

For each of the following conclusions, (1) pose a key question, (2) use a definition to provide a key answer that contains no symbols or notation, and (3) apply the key answer to the specific problem using the appropriate notation to obtain the new statement B1.

1. If m and n are odd integers, then mn is an odd integer.

   Key Question: How can I show that ______________________
   ____?

   Key Answer: Show that ____________________________________
   ____

   B1: ____

2. If the area of a right triangle XYZ is one-fourth the square of the length of the longest side, then triangle XYZ is isosceles.

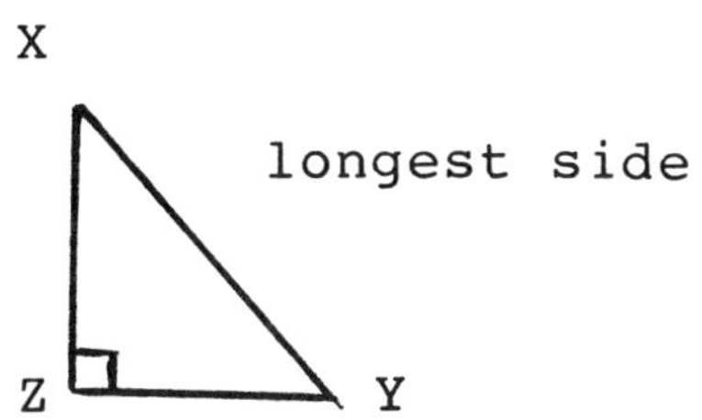

Key Question: How can I show that ______________________

____?

Key Answer: Show that ______________________

____

B1: ____

3. If the lines R, S, and T lie in the same plane, and both R and S are perpendicular to T, then R and S are parallel.

Key Question: How can I show that ______________________

____?

Key Answer: Show that ______________________

____

B1: ____

4. If s and t are rational numbers, then s + t is a rational number.

Key Question: How can I show that ____________________

____?

Key Answer: Show that ______________________________

____

B1: ____

For each of the following hypotheses, work forward by a definition to obtain a new statement A1 that uses the notation of the problem.

5. If n is an odd integer, then n^2 is an odd integer.

   A1: ____

6. If triangle XYZ is equilateral then the area of the triangle is $\sqrt{3}/4$ times the square of the length of a side.

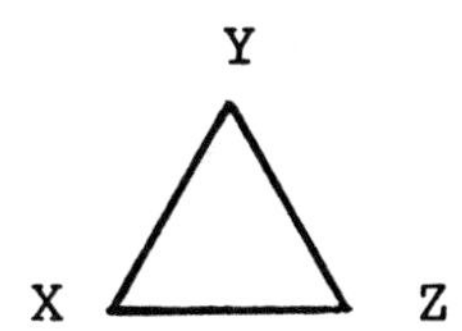

   A1: ____

7. If p is a rational number, then p^2 is a rational number.

   A1: ____

8. If n is an integer for which n^2 is even, then n is even.

   A1: ____

For each of the following proofs, fill in all of the missing steps.

9. Problem: If m and n are odd integers, then mn is an odd integer.

Analysis: The forward-backward method will be used together with the definition of an odd integer.

The Backward Process

The backward process leads to the

Key Question: How can I show that ______________________ ____?

By using the definition, one obtains the new statement

B1: mn = ________________ for some integer k.

To determine the value for k, the forward process will be used.

The Forward Process

From the hypothesis that m and n are odd integers, and by the definition of odd integers, it follows that

A1: m = ____________ for some integer p.

A2: n = ____________ for some integer q.

On multiplying the equality in A1 by that in A2, one obtains

A3: mn = _______________.

By rewriting A3, one has

A4: mn = 4pq + 2p + 2q + 1 = 2(_______) + 1.

From A4, it follows that the desired value of the integer k is ________, thus completing the proof.

Summary Table:

PROOF

STATEMENT	REASON
1. m and n are odd integers.	1. Given.
2. m = ____________ for some integer p.	2. ______________________
3. n = ____________ for some integer q.	3. ______________________
4. mn = ___________________.	4. Multiplying (2) by (3).
5. mn = 2(__________) + 1.	5. Rewriting (4).
6. mn = 2k + 1 where k = ____________, and so mn is odd.	6. ______________________

Condensed Proof: To reach the desired conclusion, it will be shown that mn = _______ for some integer k. By the hypothesis that m and n are odd integers, it follows that m = _______ for some integer p, and n = _______ for some integer q. Consequently, mn = ______________ = 4pq + 2p + 2q + 1

= 2(___________) + 1.

Therefore, the desired value for k is ___________, and mn is odd.//

10. Problem: If triangle RST is equilateral, and RS = RU, and ST = TU, then triangle RUT is equilateral.

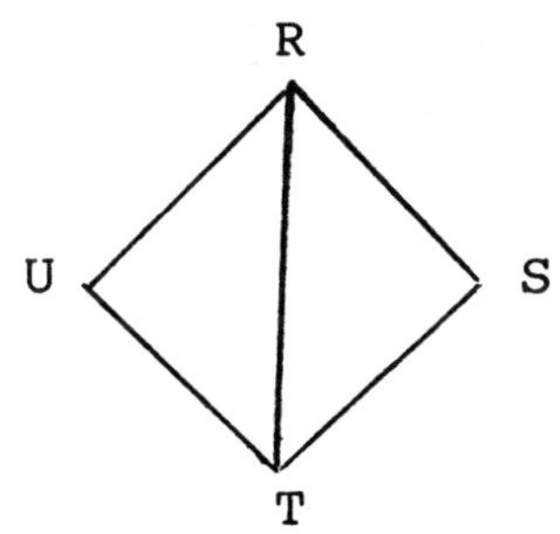

Given: Triangle RST is equilateral.

RS = RU.

ST = TU.

To Prove: Triangle RUT is equilateral.

Analysis: The forward-backward method and the definition of equilateral triangles will be used.

Summary Table:

PROOF

STATEMENT	REASON
1. ______	1. ______
2. ______	2. ______
3. ______	3. ______
4. ______	4. ______
5. ______	5. ______
6. ______	6. ______
7. ______	7. ______
8. ______	8. ______
9. ______	9. ______
10. ______	10. ______

Condensed Proof: ______

3.3 MORE ON DEFINITIONS AND TERMINOLOGY

Reversibility in Definitions (If and Only If)

When a mathematical object, say X, is defined by a statement B, you should say to yourself,

> If I have an object X, then it will satisfy the characteristics in statement B.

Reversibility means that if you come across an object satisfying the characteristics in statement B, then the object must be X. In other words, a definition works both ways. For example,

Definition: A circle C is a set of points in a plane all of whose distance from a fixed point P is the same.

If you have a set of points in a plane that is a circle, then you would know that all points in the circle are equally distant from some fixed point. By reversibility, if you have a set of points in a plane, all of which are equally distant from some fixed point, then you would know that the set is a circle.

Definitions must be reversible. To emphasize this fact, definitions are often written using the words "if and only if," as is done in the following examples:

Definition 11: Two pairs of real numbers (x_1, y_1) and (x_2, y_2) are equal if and only if $x_1 = x_2$ and $y_1 = y_2$.

Definition 12: An integer m divides an integer n if and only if there is an integer k such that n = km.

Definition 13: An integer $p > 1$ is <u>prime</u> if and only if the only positive integers that divide p are 1 and p.

*Definition 14: Two triangles are <u>similar</u> if and only if all of their corresponding angles are equal.

*Definition 15: Two triangles are <u>congruent</u> if and only if all of their corresponding sides and angles are equal.

In practice, definitions tend to have "if" instead of "if and only if." Nonetheless, it should be understood that definitions are always reversible.

<u>Equivalent Statements</u>

It is often the case that there seem to be two (or more) possible definitions for the same concept. Take, for example, the notion of an even integer that was introduced in Definition 7. A second plausible definition is:

Alternate Definition: An integer n is <u>even</u> if its remainder on division by 2 is 0.

By convention, there can be only one definition for a particular concept. Since a definition is simply something agreed on, any one of the alternatives can be agreed on as the definition. Once the definition has been chosen, it is advisable to establish the "equivalence" of the definition and the alternatives.

To establish that two statements A and B are <u>equivalent statements</u>, you must show that

1. A implies B.
2. B implies A.

When you have done so, you will know that if A is true, then B is true, and also, if B is true then so is A.

For the concept of an even integer, you use Definition 7 to create the statement

A: $n = 2k$ for some integer k

and then use the alternate definition to create the statement

B: The remainder on dividing n by 2 is 0.

Your task, then, is to prove that A implies B and that B implies A.

The statement that A is equivalent to B can be written in any of the following ways:

1. A is true if and only if B is true.
2. A if and only if B.
3. A iff B.
4. A <=> B.

More Mathematical Terminology

Given two statements A and B, you have already learned the meaning of the statement "A implies B." Several other ways of saying "A implies B" would be:

Definitions and Terminology

1. Whenever A is true, B must also be true.
2. B follows from A.
3. B is a necessary consequence of A.
4. A is sufficient for B.
5. A only if B.

You have already seen that, in order to prove that A is equivalent to B, you must show that

1. A implies B.
2. B implies A.

Statement (2) is called the <u>converse</u> of (1). For example,

Statement: If two lines are perpendicular, then they intersect in right angles. (True)

Converse: If two lines intersect in right angles, then they are perpendicular. (True)

In general, a statement and its converse need not both be true. For example,

Statement: If $n = 5$ then $n^2 = 25$. (True)

Converse: If $n^2 = 25$ then $n = 5$. (False, because it could be that $n = -5$.)

EXERCISES FOR SECTION 3.3

Indicate which of the following concepts are reversible by circling the appropriate words.

1. A triangle is a right triangle if it has a right angle. (Reversible or Not Reversible)

2. $|x| = 3$ if $x = 3$. (Reversible or Not Reversible)

3. Two angles are complementary if their degrees add up to 90°. (Reversible or Not Reversible)

4. Two angles are supplementary if they are both right angles. (Reversible or Not Reversible)

Write the reversible portions of Definitions 9 and 10. For example, the reversible portion of Definition 3 would be "If an angle has 90°, then it is a right angle."

5. Definition 9: ______________________________

   ____

6. Definition 10: ______________________________

   ____

Suppose you want to show that the following concepts are equivalent to the associated definitions. Write down the two statements that you would have to prove.

7. A triangle is equilateral if all three of its angles are equal.

   Statement 1: ______________________________

   ____

   Statement 2: ______________________________

   ____

8. Two lines in a plane are parallel if they have the same slope.

   Statement 1: ______________________________

   ____

   Statement 2: ______________________________

   ____

For each of the following statements, write the converse. Then indicate whether the original statement and its converse are true or false by circling the appropriate word.

9. If a rectangle has two adjacent sides of equal length, then the rectangle is a square. (True or False)

Converse: ______________________________________

____

______________________________________(True or False)

10. If $|x| > 4$, then $x > 4$. (True or False)

Converse: ____________________________(True or False)

11. If $x = 0$ then $xy = 0$. (True or False)

Converse: ____________________________(True or False)

12. If $\angle 1$ and $\angle 2$ are acute then they are supplementary. (True or False)

Converse: ______________________________________

______________________________________(True or False)

Suppose you want to prove the converse of the following statements. Write down what you would assume and what you would conclude.

13. If n is an integer for which n^2 is even, then n is even.

Assume: ______________________________________

Conclude: ______________________________________

14. If triangle XYZ is an isosceles triangle then $\angle X = \angle Y$.

Assume: ______________________________________

Conclude: ______________________________________

Prove the following proposition. Provide a detailed analysis of your proof together with a condensed version.

*15. If "A implies B" and "B implies C," then "A implies C."

Analysis: ______________________________

Condensed Proof: ______________________________

*16. If "A implies B," "B implies C," and "C implies A" then A is equivalent to B and A is equivalent to C.

Analysis: ______________________________

Condensed Proof: ______________________________

*17. Suppose that you have a definition in the form of a statement A together with three possible alternate definitions, say B, C, and D.

(a) How many proofs would it require to show that A is equivalent to each of the three alternatives?

____

(b) How many proofs would it require to show that "A implies B," "B implies C," "C implies D," and "D implies A?" __________

(c) Explain why the approach in part (b) is sufficient to establish that each of the alternatives is equivalent to the original definition and to each other.

____

____

____

____

3.4 USING PREVIOUS PROPOSITIONS IN PROOFS

Just as a definition can be used in the forward and backward processes, so can (previously proven) propositions, as will be shown in the next example.

Example 5: If ZXW and ZXY are right triangles, then $\angle$ ZXW and $\angle$ ZXY are supplementary.

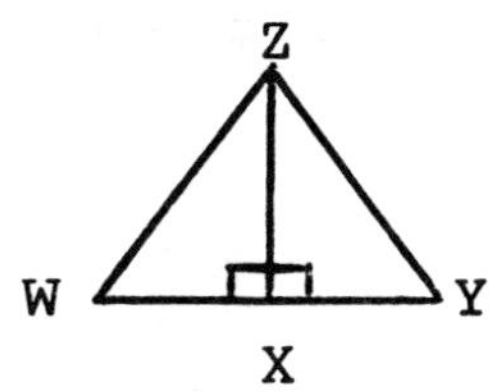

Given: Triangles ZXW and ZXY are right triangles.

To Prove: $\angle$ ZXW and $\angle$ ZXY are supplementary.

Analysis: The forward-backward method will be used to do the proof, together with the results of Example 1.

<u>The Backward Process</u>

The backward process gives rise to the

Key Question: How can I show that two angles are supplementary?

One answer can be obtained from Definition 1, but a second answer is also provided by the conclusion of Example 1, which states that $\angle$ RST and $\angle$ RSU are supplementary. Perhaps the current angles ZXW and ZXY are supplementary for the same reason as $\angle$ RST and $\angle$ RSU. To find out, the forward process will be used.

Definitions and Terminology

The Forward Process

The forward process will be used to see if $\angle$ ZXW and $\angle$ ZXY also satisfy the hypothesis of Example 1, as did $\angle$ RST and $\angle$ RSU, for then $\angle$ ZXW and $\angle$ ZXY will also satisfy the conclusion of Example 1 and hence be supplementary.

In verifying the hypothesis of Example 1 for $\angle$ ZXW and $\angle$ ZXY, it is first necessary to "match up" the current notation with that of Example 1.

Current Notation		Notation of Example 1
W	<--------------------------------->	T
X	<--------------------------------->	S
Y	<--------------------------------->	U
Z	<--------------------------------->	R
$\angle$ ZXW	<--------------------------------->	$\angle$ RST
$\angle$ ZXY	<--------------------------------->	$\angle$ RSU

Thus, to check the hypothesis of Example 1 for the current problem, you must see if $\angle$ ZXW and $\angle$ ZXY are right angles. Indeed they are, as can be established by working forward from the current hypothesis that triangles ZXW and ZXY are right triangles. By the definition of right triangles, it follows that $\angle$ ZXW and $\angle$ ZXY are right angles, and the proof is complete.

Summary Table:

PROOF

STATEMENT	REASON
1. Triangles ZXW and ZXY are right triangles.	1. Given.
2. $\angle$ ZXW and $\angle$ ZXY are right angles.	2. Definition of right triangles.
3. $\angle$ ZXW and $\angle$ ZXY are supplementary.	3. Because the hypothesis of Example 1 is true.

Condensed Proof: To reach the desired conclusion, the hypothesis of Example 1 will be verified, for then the conclusion of Example 1 will ensure that $\angle$ ZXW and $\angle$ ZXY are supplementary. But from the given fact that triangles ZXW and ZXY are right triangles, it follows that $\angle$ ZXW and $\angle$ ZXY are right angles.//

Note: Observe how much more difficult it would have been to match up the notation if the current triangles had been labeled STV and UTV. Unfortunately this "overlapping" notation can (and will) arise and, when it does, it is important to keep the symbols straight.

In summary, to use the conclusion of a previously proven proposition you should

1. Match up the notation of the current problem with that of the previous proposition.

2. Work forward from the hypothesis of the current problem to verify that the hypothesis of the previous proposition is satisfied.

3. Make use of the fact that the conclusion of the previous proposition holds for the current problem.

EXERCISES FOR SECTION 3.4

Indicate whether you would use Example 1 (see page 28) or Example 2 (see page 37) to prove the following propositions.

1. If triangle XYZ is a right triangle then the two angles that are not right angles are complementary.

    ____

2. If UV is a perpendicular bisector of WX, then ∠ UVW and ∠ UVX are supplementary.

    ____

3. If triangle WXY is a right triangle, and ∠ WYZ is a right angle, then ∠ XYW and ∠ WYZ are supplementary.

    ____

For each of the following problems, match up the notation to that of the indicated example from the text. Then, using the notation of the current problem, indicate what you would have to do in order to verify the hypothesis from the appropriate example in the text.

4. If ∠ GFI and ∠ IFE are complementary and triangle EFH is a right triangle, then ∠ EFG and ∠ EFH are supplementary.

Definitions and Terminology

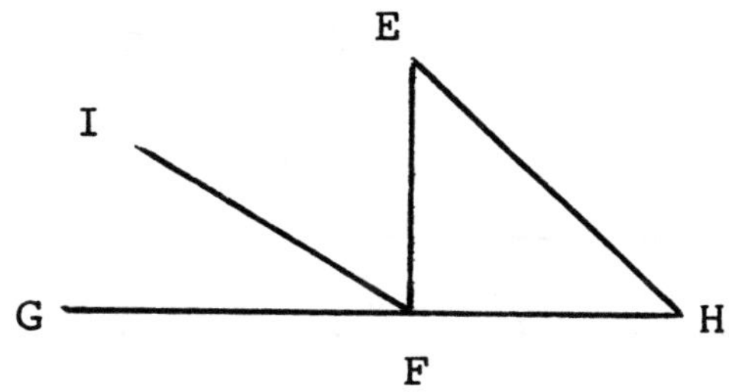

<u>Current Notation</u>   <u>Notation of Example 1</u>

_______ <-----------------------> R

_______ <-----------------------> S

_______ <-----------------------> T

_______ <-----------------------> U

_______ <-----------------------> $\angle$ RST

_______ <-----------------------> $\angle$ RSU

To verify the hypothesis of Example 1 for the current problem, you would have to show that ____________________

____

5. If triangle EFG is a right triangle in which $\angle$ EFG + $\angle$ GEF + $\angle$ FGE = 180°, then $\angle$ EFG and $\angle$ GEF are complementary.

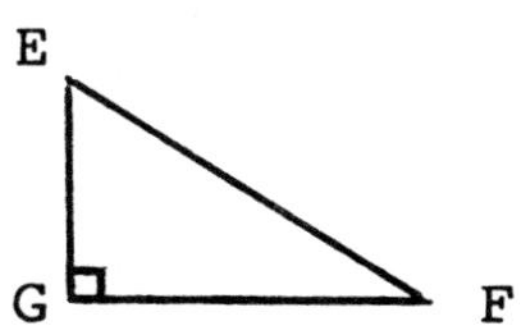

Definitions and Terminology

Current Notation		Notation of Example 2
________	<---------------------->	R
________	<---------------------->	S
________	<---------------------->	T
________	<---------------------->	$\angle 1$
________	<---------------------->	$\angle 2$
________	<---------------------->	$\angle 3$

To verify the hypothesis of Example 2 for the current problem, you would have to show that ____________________

____

*6. If the right triangle UVW satisfies $\sin(U) = \sqrt{u/2v}$, then triangle UVW is isosceles. (Note: Example 3 is on page 46.)

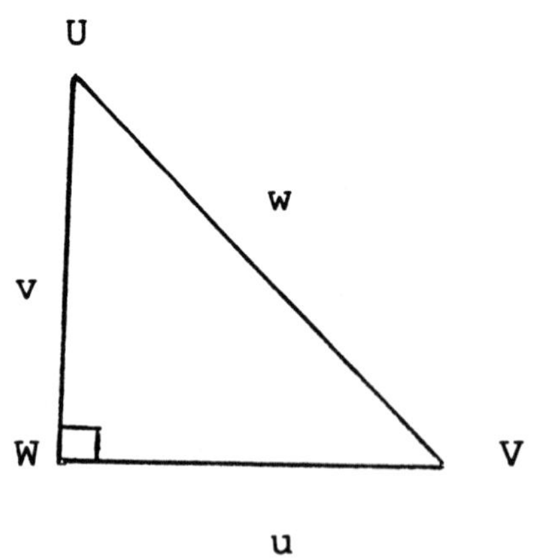

Definitions and Terminology

Current Notation	Notation of Example 3

________ <------------------------> X

________ <------------------------> Y

________ <------------------------> Z

________ <------------------------> x

________ <------------------------> y

________ <------------------------> z

To verify the hypothesis of Example 3 for the current problem, you would have to show that ____________________

____

____

Prove each of the propositions in Exercises 4, 5, and 6 by verifying the hypothesis of the appropriate example from the text. Supply a detailed analysis of the proof and a condensed version.

7. Proof of the proposition in Exercise 4.

Analysis: ______________________________

Condensed Proof: ______________________________

8. Proof of the proposition in Exercise 5.

Analysis: ______________________________

Condensed Proof: ______________________________

*9. Proof of the proposition in Exercise 6.

Analysis: ______________________________

Condensed Proof: ______________________________

CHAPTER 4
The Contradiction Method

4.1 NEGATION OF STATEMENTS

Several of the proof techniques you are about to learn require that you know what the negation of a statement is. The negation of a statement A is simply the statement that A is not true, sometimes written NOT A or, in mathematical notation, $\sim A$. In other words, NOT A is the statement that A is false. For example:

Statement A: Bill is at least as old as John.
Negation of A: It is not true that Bill is at least as old as John.

In general, the negation of a statement can be written in a simpler form. For the example above, one could write:

NOT A: Bill is not at least as old as John.

In fact, the word "not" can be eliminated altogether by incorporating it directly into the statement.

NOT A: Bill is younger than John.

To obtain the proper negation of a statement A, you should follow these two steps.

Step 1: Using statement A, write the negation in the form "It is not true that A."
Step 2: Eliminate the word "not" introduced in Step 1 by incorporating it directly into the statement.

These two steps will be illustrated with several examples.

Example: The real number x is is greater than 0.

Step 1: It is not true that the real number x is greater than 0.

Step 2: The real number x is less than or equal to 0.

Example: The integer n is even.

Step 1: It is not true that the integer n is even.

Step 2: The integer n is odd (since an integer can only be odd or even).

Example: The triangle XYZ is not equilateral.

Step 1: It is not true that the triangle XYZ is not equilateral.

Step 2: The triangle XYZ is equilateral.

Note: The occurrence of two "nots" in a statement cancel each other out.

When statement A is complicated, the incorporation of the word "not" into the statement can be tricky.

Example: The lines R and S lie in the same plane AND they do not intersect.

Step 1: It is not true that [the lines R and S lie in the same plane AND they do not intersect].

Step 2: EITHER the lines R and S do not lie in the same plane OR they do intersect.

In a similar way, when the original statement contains the words EITHER ... OR ..., its negation will contain the word AND, as is seen in the next example.

Example: EITHER $x \geq 2$ OR $x < -3$.

Step 1: It is not true that [EITHER $x \geq 2$ OR $x < -3$].

Step 2: $x < 2$ AND $x \geq -3$.

In general, the following rules hold:

Form of Statement A	Form of NOT A
1. B AND C.	1. EITHER (NOT B) OR (NOT C).
2. EITHER B OR C.	2. (NOT B) AND (NOT C).
3. NOT B.	3. B.

EXERCISES FOR SECTION 4.1

Write down the negation of each of the following statements. Be sure to incorporate the word "not" into the statement whenever possible.

1. A: n^2 is not an odd integer.

   NOT A: ______________________________

2. A: $x = y$

   NOT A: ______________________________

3. A: The set S has at most five elements.

   NOT A: ______________________________

4. A: The triangle RST is either isosceles or equilateral.

   NOT A: ______________________________

5. A: The triangle RST is neither isosceles nor equilateral.

   NOT A: ______________________________

6. A: $\angle X \neq \angle Y$ and $\angle X$ is acute.

   NOT A: ______________________________

7. A: If r and s are rational numbers, then $r + s$ is a rational number.

   NOT A: ______________________________

Reword each of the following statements so that the word "not" appears explicitly. Do not write the NOT of the statement.

Example: The statement "$x > 0$" can be rewritten as "x is not less than or equal to 0."

8. $\angle$ R is obtuse.

___

9. n is an odd integer.

___

10. Triangle XYZ is equilateral.

___

11. $x = y$.

___

12. $x < 5$ and $x > -5$.

___

13. Either $x > 5$ or $x < -5$.

___

14. There are an infinite number of even integers.

___

4.2 WHAT IS THE CONTRADICTION METHOD?

As powerful as the forward-backward method is, you may well find yourself unable to complete a proof for one reason or another, as is shown in the next example.

Example 6: If n is an integer and n^2 is even, then n is even.

Analysis: The forward-backward method will be used together with the definition of an even integer.

The Backward Process

The backward process gives rise to the

Key Question: How can I show that an integer (namely n) is even?

Using the definition of an even integer as applied to the integer n, one obtains the new statement to be proved:

B1: $n = 2k$ for some integer k.

The forward process will be used in an attempt to produce the desired integer k.

The Forward Process

Working forward from the hypothesis that n^2 is an even integer, and using the definition, results in the new statement

A1: $n^2 = 2p$ for some integer p.

The idea is to try and use p to determine the value for k. Since B1 deals with the integer n, and A1 deals with n^2, it is natural to take the square root of both sides of the equality in A1 to obtain

A2: $n = \sqrt{2p}$.

The difficulty now is that it seems impossible to rewrite $\sqrt{2p}$ in the form 2k so that the value for k can be identified. It would appear that the forward-backward method has failed!

The Contradiction Method

The contradiction method is another proof technique that you might want to try before you give up on a problem. To use the contradiction method for proving that

If A is true then B is true,

you begin by assuming that A is true, just as you would in the forward-backward method. However, to reach the desired conclusion that B is true, you ask yourself the simple question

Why can't B be false?

After all, if B is supposed to be true, then there must be some reason why B cannot be false. The objective of the contradiction method is to discover that reason. In other words, in a proof by contradiction, you

1. Assume that A is true.
2. Assume that B is not true (NOT B).
3. See why this cannot happen.

So, what does it mean to "see why this cannot happen?" Suppose, for example, that as a result of assuming that A is true and B is false, you were somehow able to reach the conclusion that 0 = 1! Surely that would convince you that it is impossible for A to be true and B to be false simultaneously. In general, to convince yourself that "this cannot happen," you work forward from statements A and NOT B in an attempt to reach a contradiction to something that you absolutely know is true. Some examples of contradictions would be:

1. $x < x$.
2. A triangle has four sides.
3. Statements C and NOT C are both true.

Thus, in a proof by contradiction, you

1. Assume that A is true.
2. Assume that NOT B is true.
3. Work forward from A and NOT B to reach a contradiction to something you know is true.

Warning: When using the contradiction method, you will not know, beforehand, what the specific contradiction is going to be. Thus, you cannot work backward. Also, each problem gives rise to its own contradiction.

Example 6 will now be proved by the contradiction method.

Example 6: If n is an integer and n^2 is even, then n is even.

Analysis: The contradiction method will be used to establish the contradiction that n^2 is both odd and even, which is impossible. By the contradiction method, you can assume:

A: n is an integer and n^2 is even.

NOT B: n is not even -- so, n is odd.

<u>Working Forward from A and NOT B</u>

Since n is odd, by the definition of an odd integer, it follows that

A1: $n = 2p + 1$ for some integer p.

To see that n^2 is an odd integer, square both sides of A1 to obtain

A2: $n^2 = (2p + 1)^2 = 4p^2 + 4p + 1$.

By rewriting A2, you have

A3: $n^2 = 2(2p^2 + 2p) + 1$.

Thus, $n^2 = 2k + 1$, where $k = 2p^2 + 2p$, and so, by definition, n^2 is odd. On the other hand, in A, it clearly states that n^2 is even. Thus, n^2 is both odd and even, and this contradiction establishes the proof.

In the condensed proof that follows, note the lack of reference to the contradiction method. Also, observe that the specific contradiction is not mentioned until the very end of the proof.

Condensed Proof: Assume that n is an odd integer. Then, by definition,

$n = 2p + 1$ for some integer p.

On squaring both sides and rewriting, one has

$$n^2 = 2(2p^2 + 2p) + 1.$$

Consequently, n^2 is an odd integer, but this contradicts the hypothesis that n^2 is even.//

It is generally the case that contradiction proofs are written in paragraph form, and so no summary table is given.

EXERCISES FOR SECTION 4.2

Write down what you would assume if you were to prove each of the following propositions by the contradiction method.

1. If the lines R and S are both perpendicular to T, then R and S do not intersect.

   Assume: ______________________________

   ____

2. If n is an integer and n^2 is odd, then n is odd.

   Assume: ______________________________

   ____

3. If $\angle R \neq \angle S$, then $\angle R$ and $\angle S$ are not vertical angles.

   Assume: ______________________________

   ____

4. If r is a real number such that $r^2 = 2$, then r is not a rational number.

   Assume: ______________________________

   ____

5. If x and y are real numbers such that $x \geq 0$, $y \geq 0$, and $x + y = 0$, then $x = 0$ and $y = 0$.

   Assume: ______________________________

   ____

Using the contradiction method, write a condensed proof for each of the following propositions. Indicate clearly what you are assuming, and what your contradiction is.

6. An integer n is not both odd and even at the same time.

   Condensed Proof: ____

7. If $x > 0$ and $y > 0$, then $x + y \neq \sqrt{x^2 + y^2}$.

   Condensed Proof: ____

8. If RS is a line, and $\angle 1$ and $\angle 3$ are complementary, then $\angle 2$ is not obtuse.

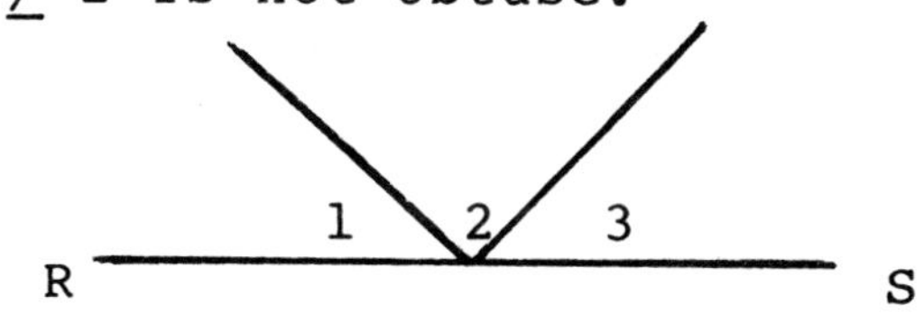

Condensed Proof: ______________________

9. If x and y are real numbers with $x \geq 0$, $y \geq 0$, and $x + y = 0$, then $x = 0$ and $y = 0$.

Condensed Proof: ______________________

10. If ∠ 1 and ∠ 2 are complementary, then ∠ 3 is acute.

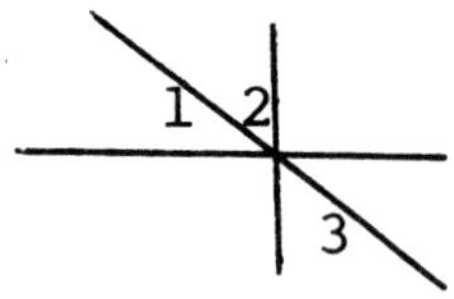

Condensed Proof: ______________________________

____

____

____

____

____

____

____

4.3 WHEN TO USE THE CONTRADICTION METHOD

A reasonable question that arises with the contradiction method is: Why and when should you use this approach instead of the forward-backward method?

The reason for doing so is that, with the forward-backward method, you can only assume that A is true, while in the contradiction method, you can assume that both A and NOT B are true. Thus, with the contradiction method, you get two statements from which to reason forward instead of just one (see Figure 3).

```
Method          Assume                                 Conclude

                       forward              backward
Forward-             A}---------> . . . <------------- B
Backward

Contra-              A\ forward
diction               }---------> . . .  *(contradiction)
                 NOT B/
```

Figure 3. The Forward-Backward Method Versus the Contradiction Method.

As a general rule, use contradiction when the statement NOT B gives you some useful information. There are at least two such recognizable instances:

1. When statement B is one of two possible cases (such as was the situtation in Example 6 where the integer n could be only odd or even).

2. When statement B contains the key word "no" or "not" in it. (This situation will be illustrated in the next example.)

The success of a proof by contradiction rests on your ability to work forward from A and NOT B so as to reach a contradiction to something that you know is true. Thus, the more statements that you know are true, the better your chances will be for reaching a contradiction. For example, if you knew the

Fact: The sum of the degrees of the three angles in any triangle is exactly 180°,

then you could easily use the contradiction method to prove the following proposition.

Example 7: If the lines R and S are both perpendicular to T, then R and S are parallel.

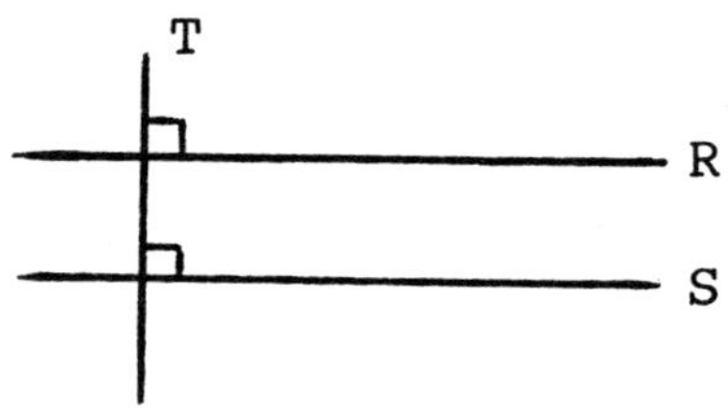

Analysis: First, the forward-backward method will be used because the conclusion of the proposition does not contain the key word "not." As you will see, when the word "not" arises in the backward process, the contradiction method will be used. At that time, a triangle whose angles add up to more than 180° will be produced, thus contradicting the given fact stated above.

The Forward-Backward Method

The conclusion of the proposition gives rise to the

Key Question: How can I show that two lines (namely R and S) are parallel?

Using Definition 10, and applying it to the lines R and S, you obtain the new statement to be proved:

B1: R and S lie in the same plane and they do not intersect.

Clearly R and S have been drawn in the same plane, so it remains only to show that

B2: R and S do not intersect.

Looking at B2, you should recognize the key word "not," and hence, you should now turn to the contradiction method.

<u>The Contradiction Method</u>

The contradiction method will now be used to show that

If the lines R and S are both perpendicular to T, then R and S do not intersect.

To do so, you should attempt to reach a contradiction by working forward from

A: The lines R and S are perpendicular to T.
NOT B: R and S do intersect.

Perhaps a picture derived from the statements A and NOT B will help you to reach a contradiction.

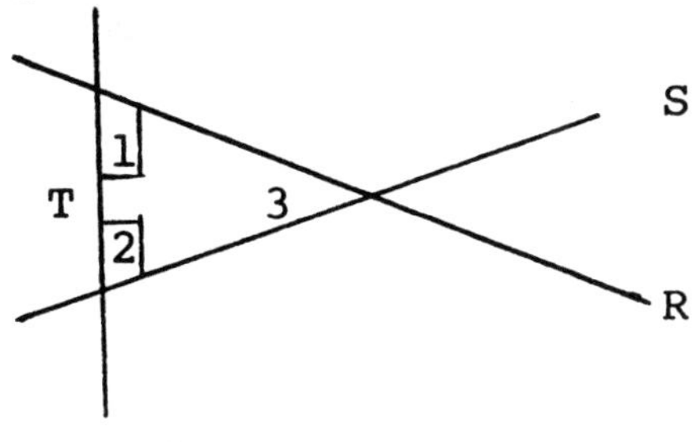

Look at the triangle in the figure. What can you say about the total degrees of the three angles? From the hypothesis that R and S are both perpendicular to T, it follows that $\angle 1 = 90^\circ$ and $\angle 2 = 90^\circ$, and therefore, $\angle 1 + \angle 2 + \angle 3 = 180^\circ + \angle 3 > 180^\circ$. In other words, the angles in the above triangle would add up to more that 180°, thus contradicting the "known" fact that the three angles of a triangle must add up to precisely 180°.

Condensed Proof: To reach the desired conclusion that R and S are parallel, observe that R and S lie in the same plane. It remains only to show that R and S do not intersect. To that end, assume that R and S do intersect.

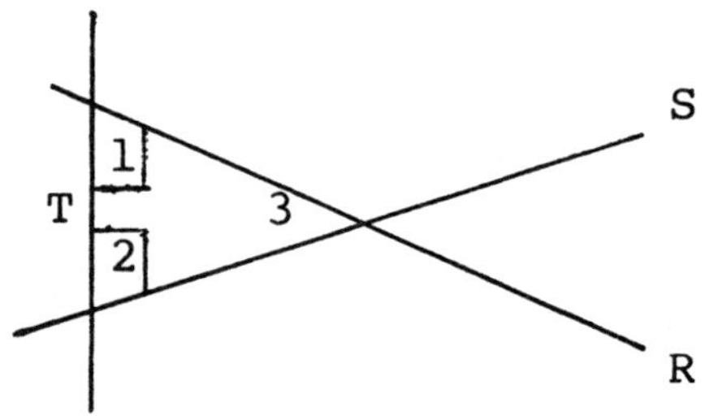

Since R and S are both perpendicular to T, $\angle 1 = \angle 2 = 90^\circ$. But then, in the triangle, $\angle 1 + \angle 2 + \angle 3 = 180^\circ + \angle 3 > 180^\circ$, contradicting the fact that the three angles of the triangle must add up to precisely 180°.//

EXERCISES FOR SECTION 4.3

For each of the following key questions, provide two key answers -- one that contains the word "not," and one that does not contain the word "not." (These exercises show that the way in which you answer a key question can determine whether you use the contradiction method or not.)

1. Key Question: How can I show that two lines are parallel?

   Key Answer: ______________________________

   ____

   Key Answer: ______________________________

   ____

2. Key Question: How can I show that an integer is even?

   Key Answer: ______________________________

   ____

   Key Answer: ______________________________

   ____

3. Key Question: How can I show that an angle is acute?

   Key Answer: ______________________________

   ____

   Key Answer: ______________________________

   ____

Contradiction

Using the contradiction method, write a condensed proof for each of the following propositions. Indicate clearly what you are assuming and what your contradiction is. You can assume the known fact that a triangle has 180°.

4. A triangle can have at most one obtuse angle.

   Condensed Proof: ______________________________

5. If $\angle 1$ and $\angle 2$ are supplementary, then R and S are parallel.

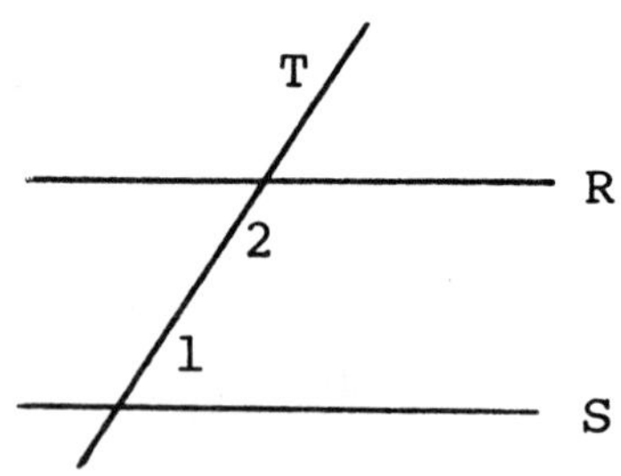

Condensed Proof: ______________________________

6. If all the angles of triangle RST are equal, but those of triangle TUV are not all equal, then RS is not parallel to UV.

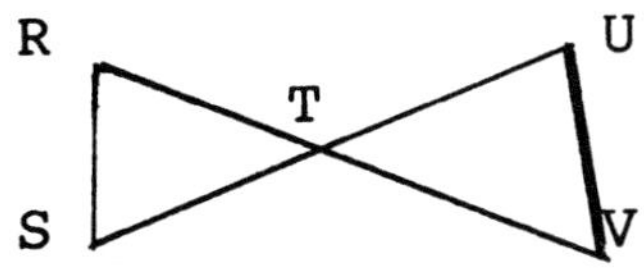

Condensed Proof: ______________________________

____

____

____

____

____

____

____

*4.4 AN ADVANCED EXAMPLE OF THE CONTRADICTION METHOD

Sometimes one has to be very creative in order to reach a contradiction, as is shown in the next example.

Example 8: If r is a real number such that $r^2 = 2$, then r is irrational.

Analysis: It is important to realize that the conclusion of the proposition can be rewritten so as to read

B: r is not rational.

In this form, the appearance of the key word "not" suggests using the contradiction method, whereby you can assume

A: r is a real number such that $r^2 = 2$.
NOT B: r is a rational number.

Working forward from NOT B, and using Definition 9 for a rational number, you can state that

A1: $r = p/q$, where p and q are integers, and $q \neq 0$.

There is still the unanswered question of where the contradiction arises. A crucial observation here will really help. It is possible to assume that p and q have no common divisor (that is, no integer that divides both p and q) other than 1 for, if they did, you could divide the common integer out of both the numerator p and the denominator q. Now, a contradiction can be reached by showing that p and q are both even, and hence 2 divides them both!

The remainder of the proof consists of working forward from A and A1 using definitions and algebraic manipulations.

Specifically, by substituting the expression for r from A1 into A, you obtain

A2: $r^2 = (p/q)^2 = p^2/q^2 = 2$

or

A3: $p^2 = 2q^2$.

Thus, p^2 is even by definition because $p^2 = 2k$ where $k = q^2$. Making use of the proposition in Example 6, since p is an integer and p^2 is even, the conclusion of Example 6 ensures that p is even.

It remains to show that q is even. Since p is even, by definition:

A4: $p = 2m$ for some integer m.

Replacing this expression for p in A3, you have

A5: $2q^2 = p^2 = (2m)^2 = 4m^2$

or

A6: $q^2 = 2m^2$.

Once again, q^2 is even because $q^2 = 2k$, where $k = m^2$. Again, by the proposition in Example 6, since q is an integer and q^2 is even, it follows that q is even, and the proof is complete.

Condensed Proof: Assume that r is a rational number of the form p/q (where p and q are integers and $q \neq 0$), and that $r^2 = 2$. Furthermore, it can be assumed that p and q have no common divisor other than 1 for, if they did, this number could be cancelled from both the numerator p and the denominator q.

Since $r^2 = 2$ and $r = p/q$, it follows that

$$r^2 = (p/q)^2 = p^2/q^2 = 2$$

or, equivalently,

$$p^2 = 2q^2.$$

Noting that $2q^2$ is even, p^2, and hence p, must be even. Consequently,

p = 2m for some integer m.

On substituting this value for p above, one has

$$2q^2 = p^2 = (2m)^2 = 4m^2$$

or, equivalently,

$$q^2 = 2m^2.$$

From this it follows that q^2, and hence q, is even. Thus, it has been shown that both p and q are even and so they have the common divisor 2. This contradiction establishes the claim.//

This particular proof was discovered in ancient times by a follower of Pythagoras, and the proof epitomizes the use of contradiction. Try to prove the proposition by some other method!

EXERCISES FOR SECTION 4.4

Using the contradiction method, write a condensed proof for each of the following propositions.

1. No chord of a circle is longer than a diameter. (Hint: You can assume that a triangle inscribed in a circle and having one side being a diameter is a right triangle.)

   Condensed Proof: ______________________________

   ____

   ____

   ____

   ____

   ____

   ____

   ____

2. If a and b are two distinct positive integers such that a divides b, then b does not divide a. (Hint: Use Definition 12 of Section 3.3.)

   Condensed Proof: ______________________________

   ____

   ____

   ____

   ____

   ____

   ____

   ____

3. There do not exist three consecutive positive integers such that the cube of the largest is equal to the sum of the cubes of the remaining two.

Condensed Proof: ______________________________

4. If triangles XYZ and VWX are not similar (see Definition 14 in Section 3.3) then VW is not parallel to YZ.

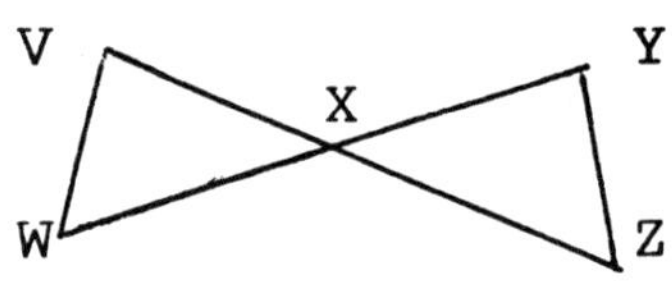

Condensed Proof: ______________________________

4.5 SUMMARY

Negation of a Statement

Given a statement A, its negation is the statement NOT A, which is the same as the statement that A is false. To find the negation of a statement A,

1. Write down the statement "It is not true that A."
2. Carefully incorporate the word "not" directly into the above statement.

Remember, a statement and its negation cannot both be true.

How to Use the Contradiction Method

When using the contradiction method to prove the proposition

If A then B,

you should

1. Assume that A is true.
2. Assume that NOT B is true.
3. Word forward from the two statements A and NOT B to reach a contradiction to something that you know is true.

When to Use the Contradiction Method

Use the contradiction method when the statement NOT B gives you some useful information. Specifically, use contradiction when

Contradiction

1. Statement B contains the key word "no" or "not."
2. Statement B refers to one of two possible cases (such as odd and even integers).
3. When the forward-backward method fails.

Summary Table

Proof Technique	When to Use It	What to Assume	What to Conclude	How to Do It
Contra-diction	When B has the key word "no" or "not" in it.	A and NOT B	Some contra-diction.	Work forward from A and NOT B to reach a contradiction.
Forward-Backward	As a first attempt, or when B does not have recognizable key words.	A	B	Work forward from A and apply the backwar[d] process to B.

CHAPTER 5
The Construction Method

5.1 WHEN TO USE THE CONSTRUCTION METHOD

In the previous chapter you learned that the contradiction method could often be used to prove a proposition in which the conclusion contains the key word "no" or "not." There are two other groups of key words, referred to as <u>quantifiers</u>, that appear repeatedly in all branches of mathematics. They are:

1. Existential quantifier -- "there is," "there are," "there exists."
2. Universal quantifier -- "for all," "for each," "for every," "for any."

The remainder of this chapter deals with the existential quantifier "there is" and the corresponding proof technique called the construction method.

The quantifier "there is" arises quite naturally in many mathematical statements. Recall Definition 7 for an even integer:

An integer n is <u>even</u> if n = 2k for some integer k.

This definition could just as well have been written using the quantifier "there is":

An integer n is <u>even</u> if there is an integer k such that n = 2k.

Another example would be the parallel postulate of geometry that states:

> Through a point P not on a line L, there is exactly one line through P that is parallel to L.

The quantifier "there is" allows for the possibility of more than one such object, as is shown in the next definition.

Definition: An integer n is a <u>square</u> if there is an integer k such that $n = k^2$.

Note: The integer 9 is square, and there are two values of k such that $n = k^2$, namely, $k = 3$ and $k = -3$.

More will be said in the next chapter about uniqueness, that is to say, the existence of only one such object.

From the examples above, you can see that each time the quantifier "there is," "there are," or "there exists" appears, the statement will have the following

Standard Form: There is an "object" with a "certain property" such that "something happens."

The words in quotation marks depend on the particular statement under consideration. Here are some examples.

1. There is an integer n greater than 2 such that $n^2 - 5n + 6 = 0$.
   Object: Integer n.
   Certain Property: $n > 2$.
   Something Happens: $n^2 - 5n + 6 = 0$.

2. Through a point P not on a line L, there is a line M through P that is perpendicular to L.
   Object: Line M.
   Certain Property: Through the point P.
   Something Happens: M is perpendicular to L.

Sometimes the quantifier is "hidden," in which case you might have to rewrite the statement in standard form so that the quantifier appears explicitly. For example,

3. Statement: An equilateral triangle contains an angle of 60°.
   Rewritten: In an equilateral triangle, there is an angle such that it measures 60°.
   Object: Angle.
   Certain Property: None.
   Something Happens: The angle measures 60°.

Mathematicians often use the symbol "$\exists$" to abbreviate the words "there is" ("there are," etc.) and the symbol "$\ni$" for the words "such that" ("for which," etc.). For example,

4. $\exists$ real numbers x and y both $> 0 \ni 2x + 3y = 8$ and $5x - y = 3$.
   Object: Real numbers x and y.
   Certain Property: $x > 0$, $y > 0$.
   Something Happens: $2x + 3y = 8$ and $5x - y = 3$.

Practice is needed to become fluent at reading and writing these statements.

EXERCISES FOR SECTION 5.1

For each of the following statements, identify the objects, the certain property, and the something that happens.

1. In the Himalayas, there is a mountain over 20,000 feet high that is taller than every other mountain in the world.

   Object: ______________________________

   Certain Property: ______________________________

   Something Happens: ______________________________

   ____

2. There exists an integer x such that $2x^2 - 5x + 3 = 0$.

   Object: ______________________________

   Certain Property: ______________________________

   Something Happens: ______________________________

   ____

3. There is a point in the interior of a square that is equally distant from the four corners of the square.

   Object: ______________________________

   Certain Property: ______________________________

   Something Happens: ______________________________

   ____

4. In an isosceles triangle, there are two acute angles with the same number of degrees.

Object: ______________________________

Certain Property: ______________________________

Something Happens: ______________________________

____

5. Between the two real numbers x and y, $\exists$ distinct rational numbers r and s $\ni |r - s| < 0.001$.

Object: ______________________________

Certain Property: ______________________________

Something Happens: ______________________________

____

6. $\exists$ an element x in the set S $\ni$ x is not in the set T.

Object: ______________________________

Certain Property: ______________________________

Something Happens: ______________________________

____

7. There is a point S in the polygon P such that, if R and T are points in P, and S is on the line segment from R to T, then R = S = T.

Object: ______________________________

Certain Property: ______________________________

Something Happens: ______________________________

____

Reword each of the following statements so that the quantifier "there is" appears explicitly in standard form.

8. In an isosceles triangle, you can find two angles with the same number of degrees.

9. A rational number r is a number that can be expressed as the ratio of two integers p and q in which the denominator q is not equal to 0.

10. At a party of n people, at least two of the people have the same number of friends at the party.

11. A triangle contains an acute angle.

*12. A polynomial of degree n, say p(x), has exactly n complex roots, say $r_1, \ldots, r_n$, for which $p(r_1) = \ldots = p(r_n) = 0$.

Construction

For each of the following statements, indicate whether the object is unique or not.

13. Through two different points, there is a line. __________

14. There is an integer x such that $2x^2 - 5x + 3 = 0$.

    ____

15. There is a real number x such that $2x^2 - 5x + 3 = 0$.

    ____

16. At the intersection of two perpendicular lines, there is a right angle. _____________

For each of the following problems, state whether you would use the forward-backward, contradiction, or construction method to begin the proof. Briefly explain your choice.

17. If n is an even integer and m is an odd integer, then n + m is an odd integer.

    Proof Technique: ____

    ____

18. If P is a point not on a line L, then you can draw a right triangle that has P as a vertex and that has one of its sides on the line L.

    Proof Technique: ____

    ____

19. If a, b, and c are numbers with $a > 0$ and $b \geq \sqrt{b^2 - 4ac}$, then there does not exist a real number $x > 0$ such that $ax^2 + bx + c = 0$.

Proof Technique: ______________________________

____

20. If R is parallel to S and S is parallel to T, then R is parallel to T.

Proof Technique: ______________________________

____

5.2 HOW TO USE THE CONSTRUCTION METHOD

During the backward process, if you ever come across a statement having the key words "there is," then one way in which you can proceed to show that the statement is true is through the construction method. The idea is to construct (guess, produce, etc.) the desired object. Of course you must show that the object you have constructed has the certain property and that the something happens.

To construct the desired object, you will generally have to use the information in the hypothesis of the proposition. Indeed, the appearance of the quantifier "there is" strongly suggests turning to the forward process to produce the desired object. Thus, to use the construction method you should

1. Turn to the forward process to construct, guess, or produce the desired object.
2. Show that the object you constructed satisfies the certain property.
3. Show that the something happens.

These steps are demonstrated with the next example.

Example 9: If the equation for a line L is given by $y = mx + b$, where m and b are real numbers and $m \neq 0$, then the line crosses the x-axis.

Analysis: On starting the backward process, you should recognize that the conclusion can be rewritten so as to read:

B: There is a value for x such that $0 = mx + b$.

Recognizing the quantifier "there is" in the standard form, it is reasonable to proceed with the construction method, whereby you should use the forward process to produce the real number x such that $0 = mx + b$.

The Forward Process

If you are clever enough to guess that the desired value for x is $-b/m$, then you are fortunate, but you must still show that the something happens, in this case, that

$$mx + b = 0.$$

Of course, for $x = -b/m$ it is true that

$$mx + b = m(-b/m) + b = -b + b = 0,$$

and thus $-b/m$ is the desired object that does satisfy the certain property and the something that happens. Also note that, by guessing this value of x, you have made use of the information in the hypothesis, since the fact that $m \neq 0$ allows you to construct $x = -b/m$.

Further Discussion

While the "guess and check" approach that was used above is perfectly acceptable for producing the desired x, it is not very informative as to how this value was produced. A more instructive proof would state that, in order for x to satisfy

$$mx + b = 0$$

it is necessary for x to satisfy

$$mx = -b.$$

Since the hypothesis states that $m \neq 0$, the last equation will hold if

$$x = -b/m.$$

Now that the construction for x has been explained, it is necessary to show that, for this value of x,

$$mx + b = 0,$$

which is of course true since

$$mx + b = m(-b/m) + b = -b + b = 0.$$

Condensed Proof: (In condensed proofs that use the construction method, the approach for constructing the desired object is often omitted.) Since the hypothesis states that $m \neq 0$, it is possible to construct $x = -b/m$. It then follows that

$$mx + b = m(-b/m) + b = -b + b = 0$$

and so the line $y = mx + b$ crosses the x-axis at $x = -b/m$.//

To be successful with the construction method, you must become a builder and use your creative ability to work forward to produce the desired object having the certain property. Also, you must not forget to show that the something happens. Your building supplies consist of the information contained in the hypothesis.

EXERCISES FOR SECTION 5.2

Explain what is wrong with each of the following proofs of the statement

There is an integer $x < 0$ such that $x^2 - x/2 - 3 = 0$.

1. Proof: The desired value for x can be found by using the quadratic formula. On so doing, the value $-3/2$ works, for then

$$x^2 - x/2 - 3 = 9/4 + 3/4 - 12/4 = 0.//$$

____

____

2. Proof: The desired value for x can be found by using the quadratic formula. On so doing, the value 2 works, for then,

$$x^2 - x/2 - 3 = 4 - 1 - 3 = 0.//$$

____

____

Complete the missing details in the following proofs. Where requested, supply a detailed analysis of your proof.

3. Problem: If r and x are rational numbers, then rx is a rational number.

Analysis: Since the conclusion does not contain a recognizable key word such as "not" or "there is," the forward-backward method will be used to begin the proof. Thus one is led to the

Key Question: How can I show that ______________________ ____?

Applying Definition 9 to the specific number rx yields the following statement to be proved:

B1: There are integers p and q with $q \neq 0$ such that $rx = p/q$.

The appearance of the quantifier "there is" in the backward process suggests using the construction method to produce the desired integers p and q. Once the integers have been constructed, it must be shown that p and q satisfy the

Certain Property: ______________________________________

Something Happens: ______________________________________

<u>The Forward Process</u>

Since the hypothesis states that r and x are rational numbers, by Definition 9, it follows that

Construction

A1: There are integers s and t with

________________ such that

$r =$ ________________.

A2: There are integers y and z with

________________ such that

$x =$ ________________.

On multiplying r and x one has

$$rx = (s/t)(y/z) = (sy)/(tz).$$

Thus the desired integers are sy for p and tz for q. To be sure that these integers satisfy the certain property and the something that happens, note that $q \neq 0$ because $t \neq 0$ and $z \neq 0$, so $q = tz \neq 0$. Also,

$$rx = (s/t)(y/z) = (sy)/(tz) = p/q.$$

Condensed Proof: ________________________________

____

____

____

____

____

____

____

4. Problem: If a and d are rational numbers, then a + d is a rational number.

Analysis: ____

____

____

____

____

____

____

____

____

____

____

____

____

____

____

Condensed Proof: ____

____

____

____

____

____

____

____

5. Problem: If P is a point not on a line L, then there is a right triangle having P as a vertex and one side on the line L.

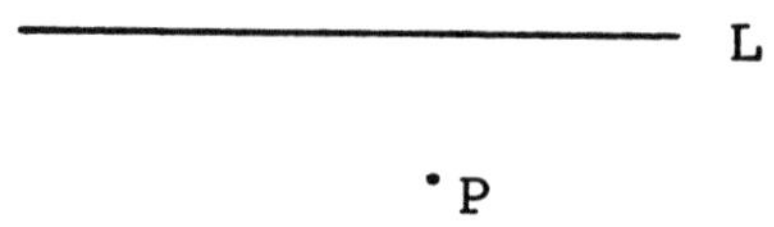

Analysis: ______________________________

____

Condensed Proof: ______________________________

____

*6. Problem: If a, b, and c are integers for which a divides b, and b divides c, then a divides c. (Hint: Use Definition 12 in Section 3.3.)

Analysis: ______________________________

Condensed Proof: ______________________________

5.3 CONSTRUCTION WITHOUT CONSTRUCTION

Another proof technique for showing that

> There is an object with a certain property such that something happens

arises from the use of contradiction. Instead of attempting to construct the desired object, why not proceed from the assumption that there is no such object? Now your job is to use this information to reach some kind of contradiction, as is illustrated in the next example.

Example 10: At a party with 367 people, there are at least two people who have the same birthday.

Analysis: If the construction method is used, then you would actually have to go out and find two such people. To save you the time and trouble, the contradiction method can be used.

In so doing, you can assume that no two people have the same birthday or, equivalently, that everyone has a different birthday. To reach a contradiction, assign numbers to the people in such a way that the person with the earliest birthday of the year receives the number 1. The person with the next earliest birthday receives the number 2, and so on.

Since each person has a different birthday, the person whose number is 1 must be born on the first day of the year (or later), the person whose number is 2 must be born on the second day of the year (or later), and so on. Consequently, the person whose number is 367 must be born on day 367 of the year (or later). But there are at most 366 days in a year, and so, a contradiction has been reached -- how can someone be born on day 367 when there are at most 366 days in the year?

This example illustrates a subtle but significant difference between a proof using the construction method and one that uses contradiction. If the construction method is successful, then you will have produced the desired object. On the other hand, if you establish the same result by contradiction, then you will know that the object exists, but you will have no way of constructing it physically. For this reason, it is often the case that proofs done by contradiction are shorter and easier than those done by construction because, with contradiction, you do not have to create the desired object. You only have to show that its nonexistence is impossible! This difference has led to some great philosophical debates throughout the ages.

In summary, to use the contradiction method in this type of proof, you should

1. Assume that there is no object with the certain property such that the something happens.
2. Work forward from this information and the hypothesis to reach a contradiction.

EXERCISES FOR SECTION 5.3

Prove each of the following propositions by using the contradiction method.

1. There are at least two people out of 4,000,000,000 people in the world with the same number of hairs on their heads. (Hint: Use the same kind of argument as that of Example 10, keeping in mind that no one has more than 250,000 hairs on their heads.)

   Condensed Proof: ____

2. In a party of n ($n \geq 2$) people, there are at least two people who have the same number of friends at the party. (Hint: Use the same kind of argument as that of Example 10, and look carefully at the person with the fewest friends and the person with the most friends.)

Construction

Condensed Proof: ______________________________

____

____

____

____

____

____

____

3. If you were to use the contradiction method to prove that

> There is exactly one object with a certain property such that something happens,

then you would assume that the above statement is not true. In that case, how many objects will there be that satisfy the certain property and the something that happens? ______________________________

5.4 SUMMARY

When to Use the Construction Method

Use the construction method when the quantifier "there is" ("there are," "there exist," etc.) arises in the backward process in the

Standard Form: There is an object with a certain property such that something happens.

Remember to watch for hidden quantifiers and to rewrite such statements when necessary.

How to Use the Construction Method

To do a proof with the construction method,

1. Use the forward process to construct, guess, or produce the desired object.
2. Show that the object thus constructed satisfies the certain property.
3. Show that the object also satisfies the something that happens.

Using Contradiction to Show Existence

To use the contradiction method for showing that there is an object with a certain property such that something happens, proceed by

1. Assuming that there is no such object.
2. Working forward from this fact and the hypothesis to reach a contradiction.

Summary Table

Proof Technique	When to Use It	What to Assume	What to Conclude	How to Do It
Construction	When B has the key words "there is," "there are," etc.	A	There is the desired object.	Guess, construct, etc. the object having the certain property, and show that the something happens.
Contradiction	When B has the key word "no" or "not" in it.	A and NOT B	Some contradiction.	Work forward from A and NOT B to reach a contradiction.

CHAPTER 6
The Uniqueness Methods

6.1 WHEN TO USE THE UNIQUENESS METHODS

As with most proof techniques, the time to use the uniqueness methods is when certain key words arise. In the previous chapter, you learned to use the construction method when, in the backward process, you come across the key words "there is" appearing in the standard form:

> There is an object with a certain property such that something happens.

In such problems, it is often the case that the desired object is unique, that is to say, there is only one object that satisfies the certain property and the something that happens.

You should use a uniqueness method when the key word "unique" arises in the backward process in the

Standard Form: There is a unique object with a certain property such that something happens.

Other key words that are equivalent to the concept of uniqueness are:

1. One and only one.
2. Exactly one.
3. Precisely one.
4. Determine.

Uniqueness

Each of the following statements deals with the issue of uniqueness.

1. Through a point P not on a line L, there is a unique line that is parallel to L.
2. For a given real number r, there is one and only one real number x such that $x^3 = r$.
3. If two different lines intersect, then they intersect in exactly one point.
4. A line segment has precisely one midpoint.
5. Three points not all on the same line determine a plane.

Whenever any key word equivalent to that of "uniqueness" arises in the backward process, the uniqueness methods, to be described in this chapter, are likely to be successful.

EXERCISES FOR SECTION 6.1

Reword each of the following statements so that the word "unique" appears in standard form.

1. Two distinct points determine a line.

____

____

2. Through a point P not on a line L, there is exactly one line that is perpendicular to L.

____

____

3. A right triangle contains precisely one right angle.

____

____

4. If a, b, c, d, e, and f are real numbers with $ad - bc \neq 0$, then there is one and only one solution to the system of equations

$$ax + by = e$$
$$cx + dy = f.$$

____

____

____

Uniqueness

For each of the following existence statements, indicate whether the object is unique or not.

5. There is a real number x with $x > 0$ such that $x^2 - 5x + 6 = 0$. ____________________

6. If two different planes intersect, then they intersect in a line. ____________________

7. An acute angle X has a complementary angle Y. __________

8. A square has a diagonal. ____________________

9. If P is a point not on a line L, then there is a right triangle having P as a vertex, and one of its sides on the line L. ____________________

6.2 THE DIRECT UNIQUENESS METHOD

When faced with the issue of trying to prove that there is a unique object with a certain property such that something happens, the very first step is to establish that, indeed, there is an object with the certain property such that the something happens. As you learned in the previous chapter, this task could be accomplished by using either the construction or the contradiction method. Then the issue of uniqueness must be addressed.

One approach for proving uniqueness comes from the following idea. Suppose, for the moment, that both X and Y are objects that satisfy the certain property and the something that happens. In this event, in order to convince yourself that there really is only one such object, it had better be the case that X and Y are really the same, that is, X = Y.

In other words, when you want to prove that

> If A is true, then there is a unique object with a certain property such that something happens,

you should

1. Establish the existence of such an object by using the construction or contradiction method.
2. Assume that X is an object with the certain property such that the something happens.
3. Assume that Y is an object with the certain property such that the something happens.
4. Work forward from (2), (3), and the hypothesis A to establish that X and Y are the same, that is, that X = Y.

Note: In (4), you can work backward to establish that X = Y.

This direct uniqueness method will now be illustrated.

Example 11: If the line L is represented by the equation $y = mx + b$, and $m \neq 0$, then the line L crosses the x-axis in exactly one point.

Analysis: Recognizing the key words "exactly one" in the conclusion of the proposition, you should consider using the direct uniqueness method to show that

B: There is a unique value of x for which $0 = mx + b$.

According to the discussion above, the first step would be to establish the existence of a value for x that satisfies $0 = mx + b$. Since this task was already accomplished in Example 9 of the previous chapter, there is no need to repeat the argument here.

To address the issue of uniqueness, you should assume that

A1: x is a real number that satisfies $0 = mx + b$.
A2: w is a real number that satisfies $0 = mw + b$.

The objective is to work forward from A1 and A2, and from the hypothesis that $m \neq 0$, to establish that

B1: $x = w$.

To do so, combine A1 and A2 to obtain

A3: $mx + b = 0 = mw + b$.

Then, by subtracting b from both sides of A3, you have

A4: $mx = mw$.

Now you can use the hypothesis that $m \neq 0$ to divide both sides of A4 by m to obtain precisely

B1: $x = w$,

thus completing the proof.

Condensed Proof: The existence of the real number x satisfying $0 = mx + b$ was already established in Example 9, and so those details will be omitted. Suppose now that both x and w satisfy

$$0 = mx + b$$
$$0 = mw + b.$$

(It is this statement that indicates that the direct uniqueness method is being used.) Then, by algebra, it follows that

$$mx = mw.$$

Finally, since $m \neq 0$ by hypothesis,

$$x = w$$

and so the uniqueness has been proved.//

EXERCISES FOR SECTION 6.2

Fill in the missing details for the following proof.

1. Proposition: If $x > 2$, then there is a unique value of y satisfying $y < 0$ such that $x = (2 - x)y$.

   Analysis: The key word "unique" in the conclusion suggests using the direct uniqueness method. Thus, the desired value for y must first be constructed. Indeed, since $x > 2$, $2 - x \neq 0$ and so it is possible to construct y, namely

   $$y = x/(2 - x).$$

   To be sure that this value of y is correct, it is necessary to show that

   Certain Property: ______________________________

   Something Happens: ______________________________

   But since by hypothesis $x > 2$, the numerator x in the expression for y is greater than 0 while the denominator $(2 - x)$ is less than 0, and thus $y < 0$. Also, to check that the something happens, it is easy to verify that

   $$(2 - x)y = (2 - x)[x/(2 - x)] = x.$$

Uniqueness

To address the issue of uniqueness, suppose that both y and z satisfy

A1: ____________ and x = ________________

A2: ____________ and x = ________________

Working forward from A1 and A2 together with the hypothesis that $x > 2$, it must be shown that

B1: ______________________________________

Combining A1 and A2 it follows that

A3: $(2 - x)y = (2 - x)z$.

Finally, from the hypothesis that $x > 2$, $2 - x \neq 0$ and so one can divide both sides of A3 by _____________ to obtain precisely B1.

2. Explain what is wrong with the following uniqueness proof.

Proposition: If $\angle 1$ is a right angle, then there is a unique angle (in terms of its measure) that is supplementary to $\angle 1$.

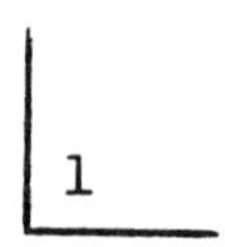

Proof: Suppose that $\angle 2$ and $\angle 3$ are both supplementary to $\angle 1$. It will be shown that $\angle 2 = \angle 3$. By the hypothesis that $\angle 1$ is a right angle, it follows that $\angle 1 = 90^\circ$. Also, since $\angle 1$ and $\angle 2$ are supplementary, $\angle 1 + \angle 2 = 180^\circ$, so $90^\circ + \angle 2 = 180^\circ$, or $\angle 2 = 90^\circ$. Similarly, since $\angle 1$ and $\angle 3$ are supplementary, $\angle 1 + \angle 3 = 180^\circ$, so $90^\circ + \angle 3 = 180^\circ$, so $\angle 3 = 90^\circ$. Thus $\angle 2 = \angle 3$ and the proof is complete.//

Use the direct uniqueness method to prove each of the following propositions. Supply a detailed analysis together with a condensed proof. Briefly address the issue of construction. Spend most of your efforts on establishing the uniqueness.

3. If ∠ X is acute, then there is a unique ∠ Y (in terms of its measure) such that ∠ X and ∠ Y are complementary.

Analysis: ______________________________

Condensed Proof: ______________________________

4. If y is a real number greater than 0, then there is a unique value of x satisfying $x > 0$ such that $xy = 2$.

Analysis: ______________________________

Condensed Proof: ______________________________

*5. If a, b, c, d, e, and f are real numbers with $ad - bc \neq 0$, then there is a unique solution to the equations

$$ax + by = e$$
$$cx + dy = f.$$

Analysis: ______________________________

Condensed Proof: ______________________________

*6. If a and b are real numbers, at least one of which is not 0, and $i = \sqrt{-1}$, then there is a unique complex number, say $c + di$, such that $(a + bi)(c + di) = 1$.

Analysis: ______________________________

Condensed Proof: ______________________________

6.3 THE INDIRECT UNIQUENESS METHOD

A second method, based on the principle of contradiction, is available for trying to prove that there is a unique object with a certain property such that something happens. As with the direct uniqueness method, the first step is to establish the existence of the object.

Then, to prove the uniqueness, the approach of the indirect method is slightly different. You assume that X and Y are two different objects with the certain property and for which the something happens. Now, in theory, these two objects cannot exist because there is only supposed to be one such object. In order to convince yourself that it is impossible for both X and Y to exist, you must reach a contradiction.

In other words, when you want to prove, by the indirect uniqueness method, that

> If A is true, then there is a unique object with a certain property such that something happens,

you should

1. Establish the existence of such an object by using the construction or contradiction method.
2. Assume that X is an object with the certain property such that the something happens.
3. Assume that Y is an object with the certain property such that the something happens.
4. Assume that X and Y are different, that is, that $X \neq Y$.
5. Work forward from (2), (3), and (4), together with the hypothesis A to establish a contradiction.

Note: You cannot work backward because you will not know, beforehand, what the contradiction is going to be.

The indirect method for establishing uniqueness will now be illustrated.

Example 12: If L and M are two different lines that intersect, then they intersect in one and only one point.

Figure:

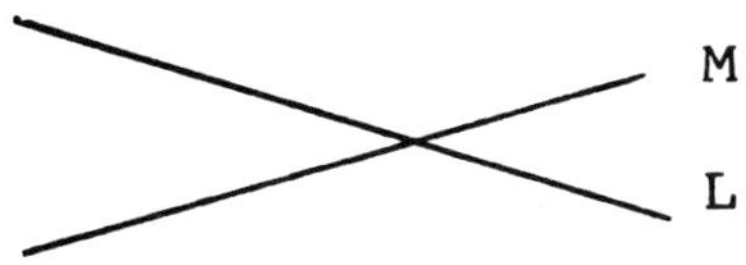

Analysis: Recognizing the key words "one and only one" you should consider using a uniqueness method. Here, the indirect uniqueness method will be used to show that

B: The lines L and M intersect in a unique point.

According to the discussion above, the first step is to show the existence of a point of intersection for the lines L and M. In this problem, the hypothesis explicitly states that the two lines intersect.

To address the issue of uniqueness by the indirect method, you should assume that

A1: R is a point of intersection for L and M.
A2: S is a point of intersection for L and M.
A3: R and S are different points.

The objective is to work forward from A1, A2, A3, and the hypothesis, if necessary, to reach a contradiction. In this case, a contradiction can be reached by using the postulate:

P1: Between two different points, there is one and only one line.

Specifically, from A3, R and S are two different points, so by P1, there should be only one line through R and S. Yet A1, A2, and the hypothesis indicate that L and M are two different lines containing R and S, thus completing the proof.

Condensed Proof: (Note that nowhere will it say that the indirect uniqueness method is being used. Moreover, the contradiction is not stated until the very end.) The hypothesis states that L and M do intersect. To show the uniqueness, suppose that R and S are two different points of intersection of the lines L and M. But then, L and M would be two different lines containing the distinct points R and S, which contradicts the postulate that between two different points, there is one and only one line.//

Remember that, in order to reach a contradiction with the indirect uniqueness method, you will have to make use of the fact that the two objects are different.

EXERCISES FOR SECTION 6.3

Use the indirect uniqueness method to prove each of the following propositions. Supply a detailed analysis and a condensed proof. Briefly address the issue of construction. Spend most of your efforts on establishing uniqueness.

1. If the equation for a line L is given by $y = mx + b$, and $m \neq 0$, then the line crosses the x-axis in a unique point.

   Analysis: ______________________________

Condensed Proof: ______________________________

2. If $\angle X < 180^\circ$, then there is a unique $\angle Y$ (in terms of its measure) such that $\angle X$ and $\angle Y$ are supplementary.

Analysis: ______________________________

Condensed Proof: ______________________________

3. There is a unique integer x satisfying $x > 0$ such that $x^2 = 25$.

Analysis: ______________________________

Condensed Proof: ____

4. In a right triangle, there is a unique right angle. (Hint: You can use the fact that the three angles of a triangle add up to 180°.)

Analysis: ____

Condensed Proof: ______________________________

____

____

____

____

____

____

____

*5. If $r > 0$, then there is a unique real number x such that $x^3 = r$. (Note: Do not establish existence. Prove only the uniqueness. To do so, assume that x and y are two different numbers satisfying $x^3 = r$ and $y^3 = r$. Then factor $x^3 - y^3$ into $(x^2 + xy + y^2)(x - y)$, and then think of $x^2 + xy + y^2$ as a quadratic equation of the form $ax^2 + bx + c$ where $a = 1$, $b = y$, $c = y^2$.)

Analysis: ______________________________

____

____

____

____

____

____

____

____

____

Uniqueness

Condensed Proof:

6.4 SUMMARY

When to Use the Uniqueness Methods

Use the uniqueness methods when you come across any of the key words:

1. Unique.
2. One and only one.
3. Exactly one.
4. Precisely one.
5. Determine.

In general, such a problem will have the form:

If A is true, then there is a unique object with a certain property such that something happens.

The Direct Uniqueness Method

To use this method, you should

1. Establish the existence of such an object by using the construction or contradiction method.
2. Assume that X and Y are two objects with the certain property and for which the something happens.
3. Work forward from (2) and the hypothesis A to establish that X and Y are the same, that is, that X = Y.

Note: You can, of course, work backward from X = Y.

The Indirect Uniqueness Method

To use this method, you should

1. Establish the existence of such an object by using the construction or contradiction method.
2. Assume that X and Y are two <u>different</u> objects that satisfy the certain property and the something that happens.
3. Work forward from (2) and the hypothesis A to reach a contradiction.

Note: You cannot work backward because you will not know, beforehand, what the contradiction is going to be.

Summary Table

Proof Technique	When to Use It	What to Assume	What to Conclude	How to Do It
Uniqueness (Direct)	When B has the key word "unique" in it.	X and Y are two such objects, and A.	The two objects are equal.	Work forward from A and the properties of X and Y. Also work backward to show that X = Y.
Uniqueness (Indirect)	When B has the key word "unique" in it.	X and Y are two different objects, and A.	Some contradiction.	Work forward from A, the properties of X and Y, and the fact that X and Y are different, to reach a contradiction.
Construction	When B has the key words "there is," "there are," etc.	A	There is the desired object.	Guess, construct, etc the object having the certain property, and show that the something happens.

CHAPTER 7
The Contrapositive Method

7.1 HOW TO USE THE CONTRAPOSITIVE METHOD

In Chapter 4, the contradiction method was presented as one technique for proving that

If A is true then B is true.

There, you learned to work forward from the two statements A and NOT B to reach some kind of contradiction. In general, the difficulty with this method is that you do not know what the contradiction is going to be. As will be seen in this chapter, the contrapositive method has the advantage of directing you toward one specific type of contradiction.

The contrapositive method is similar to contradiction in that you begin by assuming that A and NOT B are true. Unlike contradiction, however, you do not work forward from both A and NOT B. Instead, you work forward only from NOT B. Your objective is to reach the contradiction that A is false (written NOT A), for how can A be true and false at the same time?

Simply stated, with the contrapositive method you should

1. Write down the statements NOT B and NOT A.
2. Assume that A and NOT B are true.
3. Work forward from NOT B to obtain NOT A.

Note 1: You can work backward from NOT A.

Note 2: Since statement A is not actively being used, there is no need to state explicitly that you are assuming A is true.

The next example illustrates the contrapositive method.

Example 13: If p and q are real numbers such that $pq \neq (p + q)^2/4$, then $p \neq q$.

Analysis: In this proposition, one has the statements

NOT B: $p = q$.
NOT A: $pq = (p + q)^2/4$.

According to the contrapositive method, the objective is to work forward from NOT B to obtain NOT A. Since one can work backward from NOT A, you can ask the

Key Question: How can I show that two numbers (namely pq and $(p + q)^2/4$) are equal?

In this case, the answer is to show that both numbers are equal to p^2, that is, that

B1: $pq = p^2$

and

B2: $(p + q)^2/4 = p^2$.

B1 can be obtained immediately by multiplying both sides of the equality in NOT B by p. Finally, to see that B2 is true, it follows from NOT B that

$$(p + q)^2/4 = (p + p)^2/4 = (2p)^2/4 = 4p^2/4 = p^2$$

and the proof is complete.

Condensed Proof: (Note the lack of reference to the contrapositive method and to the fact that the statement NOT A has been reached.) Assume that $p = q$. But then it follows that

$$p^2 = pp = pq.$$

Also,

$$(p + q)^2/4 = (p + p)^2/4 = p^2$$

and so

$$pq = (p + q)^2/4.//$$

Observe that, since the contrapositive method is a type of proof by contradiction, the contrapositive method should be considered seriously when the conclusion of the proposition contains the key word "no" or "not." It is often preferable to use the contrapositive method before contradiction because you will know what you are looking for, and hence you can work backward.

EXERCISES FOR SECTION 7.1

When using the contrapositive method for proving each of the following propositions, indicate which statement(s) you would use to work forward and which statement you would use to work backward.

1. If n is an integer for which n^2 is even, then n is an even integer.

   Forward: ____

   Backward: ____

2. If x and y are real numbers with $x \neq y$, then $f(x) \neq f(y)$ (where f is a function).

   Forward: ____

   Backward: ____

3. If no angle of a quadrilateral RSTU is obtuse, then the quadrilateral RSTU is a rectangle.

   Forward: ____

   Backward: ____

4. If c is an odd integer, then the equation $n^2 + n = c$ has no odd integer solution for n.

   Forward: ____

   Backward: ____

Contrapositive

Circle the correct choice for each of the following problems.

5. When the contrapositive method is used to prove the proposition "If the lines R and S are not parallel, then $\angle 1 \neq \angle 2$," then which of the following is the correct key question?

   (a) How can I show that $\angle 1 = \angle 2$?

   (b) How can I show that R and S are parallel?

   (c) How can I show that two angles are equal?

   (d) How can I show that two lines are parallel?

6. If the contrapositive method is used to prove the proposition "If c is a real number less than -1, then there does not exist a real number x between 0 and 2 such that $x^2 + 2x \leq 2c$," then which of the following is the result of the forward process?

   (a) $c \geq -1$.

   (b) $x^2 \leq 2(c - x)$.

   (c) $c + 1 < 0$.

   (d) $x^2 + 2x \geq 2c$.

Prove each of the following propositions by the contrapositive method. Supply a detailed analysis and a condensed proof.

7. Suppose that m and b are real numbers with $m \neq 0$ and let L be the line defined by $y = mx + b$. Prove that if $x \neq w$, then $mx + b \neq mw + b$.

Analysis: ______________________________

Condensed Proof: ______________________________

8. If c is an odd integer, then the equation $n^2 + n = c$ has no odd integer solution for n.

Analysis: ______________________________

Condensed Proof: ______________________________

7.2 CONTRAPOSITIVE, INVERSE, AND TRUTH TABLES

Contrapositive and Inverse Statement

You have already learned the meaning of the

Statement: A implies B

and you have also seen the

Converse: B implies A.

Two other related statements of interest are:

Contrapositive: NOT B implies NOT A.
Inverse: NOT A implies NOT B.

All of these statements are illustrated in the next example.

Statement:	If $\angle 1$ and $\angle 2$ are complementary, then $\angle 1 + \angle 2 = 90^\circ$. (True)
Converse:	If $\angle 1 + \angle 2 = 90^\circ$, then $\angle 1$ and $\angle 2$ are complementary. (True)
Contrapositive:	If $\angle 1 + \angle 2 \neq 90^\circ$, then $\angle 1$ and $\angle 2$ are not complementary. (True)
Inverse:	If $\angle 1$ and $\angle 2$ are not complementary, then $\angle 1 + \angle 2 \neq 90^\circ$. (True)

You might recall from Chapter 3 that a statement and its converse need not both be true. Similarly, the contrapositive and inverse statements need not both be true simultaneously, as is demonstrated in the next example.

Statement:	If ∠ 1 and ∠ 2 are right angles, then ∠ 1 and ∠ 2 are supplementary. (True)
Converse:	If ∠ 1 and ∠ 2 are supplementary, then ∠ 1 and ∠ 2 are right angles. (False)
Contrapositive:	If ∠ 1 and ∠ 2 are not supplementary, then either ∠ 1 or ∠ 2 is not a right angle. (True)
Inverse:	If either ∠ 1 or ∠ 2 is not a right angle, then ∠ 1 and ∠ 2 are not supplementary. (False)

It is a fact that

1. The original statement and the contrapositive are both true or both false.
2. The converse and inverse statements are both true or both false.

*Truth Tables

The truth table for the statement

A implies B

(see Table 1 in Section 1.3) can be used to determine when the converse, contrapositive, and inverse statements are true. For instance, the contrapositive statement

NOT B implies NOT A

is true in all cases except when the statement to the left of the word "implies" (namely NOT B) is true and the statement to the right of the word "implies" (namely NOT A) is false. In other words, the contrapositive statement is true in all cases except when B is false and A is true, as is shown in Table 4.

Table 4. The Truth Table for "NOT B implies NOT A" (T = True, F = False)

****

A	B	NOT B	NOT A	A => B	NOT B => NOT A
T	T	F	F	T	T
T	F	T	F	F	F
F	T	F	T	T	T
F	F	T	T	T	T

****

In a similar manner, truth tables can be derived for the converse and inverse statements, and these will be left to the exercises.

EXERCISES FOR SECTION 7.2

For each of the following statements, write the converse, contrapositive, and inverse statements. Indicate whether each statement is true or false by circling the appropriate word.

1. Statement: If a triangle is a right triangle, then it has two complementary angles. (True or False)

   Converse: ________________________________

   ____

   ______________________________ (True or False)

   Contrapositive: ___________________________

   ____

   ______________________________ (True or False)

   Inverse: _________________________________

   ____

   ______________________________ (True or False)

2. Statement: (R, S, and T are lines with T intersecting both R and S.) If R is parallel to S, then R is perpendicular to T and S is perpendicular to T. (True or False)

   Converse: ________________________________

   ____

   ______________________________ (True or False)

Contrapositive: ______________________________________

____

________________________________ (True or False)

Inverse: ____

____

________________________________ (True or False)

3. Statement: If the quadrilateral WXYZ is a parallelogram with at least one right angle, then the quadrilateral WXYZ is a rectangle. (True or False)

Converse: ____

____

________________________________ (True or False)

Contrapositive: ______________________________________

____

________________________________ (True or False)

Inverse: ____

____

________________________________ (True or False)

Fill in the following truth tables.

*4. Statement: NOT A implies NOT B.

A	B	NOT A	NOT B	NOT A => NOT B
T	T	__	__	__
T	F	__	__	__
F	T	__	__	__
F	F	__	__	__

*5. Statement: (NOT A) OR B.

A	B	NOT A	(NOT A) OR B
T	T	___	___
T	F	___	___
F	T	___	___
F	F	___	___

*6. How is the truth table in the previous exercise related to that of "A implies B?"

____

____

____

7.3 SUMMARY AND COMPARISON OF THE CONTRAPOSITIVE METHOD

Contrapositive Versus Contradiction

As was mentioned in the previous section, one of the primary advantages of the contrapositive method (over contradiction) is that you know precisely what statement you are trying to reach, namely NOT A. As such, you can apply the backward process in the contrapositive method. That option is not available in the contradiction method.

On the other hand, there is a disadvantage to the contrapositive method, that being, that you work forward only from NOT B. In contrast, with contradiction, you can work forward from both A and NOT B. These advantages and disadvantages have been summarized in Figure 4.

```
Method          Assume                                    Conclude

Contra-            A
positive               forward              backward
                NOT B}---------> . . . <------------ NOT A

Contra-            A\ forward
diction             }---------> . . .  *  (contradiction)
                NOT B/
```

Figure 4. The Contrapositive Method Versus the Contradiction Method.

Contrapositive Versus Forward-Backward

In both the contrapositive and the forward-backward methods, you work forward from one statement and backward from one statement. It is natural to ask why and when one method might be better than the other. The answer comes from recognizing that the two methods work forward and backward from different statements! In the forward-backward method you work forward from A and backward from B. In the contrapositive method, however, you work forward from NOT B and backward from NOT A. Perhaps it is better to work forward from NOT B than from A. Similarly, it might be easier to apply the backward process to NOT A than to B. As a general rule, use the contrapositive method when the conclusion of the proposition contains the word "no" or "not." These observations have been summarized in Figure 5.

```
Method          Assume                                  Conclude

Forward-             forward             backward
Backward           A}---------> . . . <---------------- B


Contra-            A
positive             forward             backward
               NOT B}---------> . . . <------------ NOT A
```

Figure 5. The Forward-Backward Method Versus the Contrapositive Method.

Forward-Backward Versus Contradiction Versus Contrapositive

The advantages and disadvantages of all three methods can be seen from Figure 6.

```
Method          Assume                                  Conclude

Forward-             forward             backward
Backward           A}---------> . . . <---------------- B

Contra-            A} forward
diction             }---------> . . .  *  (contradiction)
               NOT B}

Contra-            A
positive             forward             backward
               NOT B}---------> . . . <------------ NOT A
```

Figure 6. The Forward-Backward Method Versus the Contradiction Method Versus the Contrapositive Method.

How to Use the Contrapositive Method

To apply the contrapositive method to the proposition

If A is true then B is true,

you should

1. Write down the statements NOT B and NOT A.
2. Assume that NOT B is true.
3. Work forward from NOT B to reach NOT A.

Note 1: You can work backward from NOT A.

Note 2: Use the contrapositive method (even before contradiction) when B contains the word "no" or "not."

Summary Table

Proof Technique	When to Use It	What to Assume	What to Conclude	How to Do It
Contra-positive	When B has the key word "no" or "not" in it.	NOT B	NOT A	Work forward from NOT B and backward from NOT A.
Contra-diction	When B has the key word "no" or "not" in it.	A and NOT B	Some contra-diction.	Work forward from A and NOT B to reach a contradiction.
Forward-Backward	As a first attempt, or when B does not have recognizable key words.	A	B	Work forward from A and apply the backward process to B.

CHAPTER 8
Putting It All Together

8.1 SUMMARY OF THE PROOF TECHNIQUES

In this section, a review of the proof techniques for proving that

If A is true then B is true

will be given.

The Forward-Backward Method

With the forward-backward method you can assume that A is true and, by using this assumption, your objective is to reach the conclusion that B is true. Through the forward process, you derive from A, a sequence of statements A1, A2, ... that are true as a result of A being true. The sequence is guided by the backward process whereby, through asking and answering the key question, you derive from B, a new statement, B1, with the property that if B1 is true, then so is B. The backward process can then be applied to B1, obtaining a new statement, B2, and so on. The objective is to link the forward sequence to the backward sequence by generating a statement in the forward sequence that is precisely the same as the last statement obtained in the backward sequence.

Do not forget the role of definitions in both the forward and backward processes. Also, as you generate statements in the backward process, watch for certain key words to appear, for then the associated proof technique is likely to be successful. As a general rule, use the forward-backward method when the conclusion contains no recognizable key words.

The Contradiction Method

With the contradiction method, you try to establish that B is true by asking yourself why B cannot be false. Specifically, you should

1. Assume that A and NOT B are true.
2. Work forward from A and NOT B to reach a contradiction to something you know is true.

You should consider using the contradiction method when B contains the key word "no" or "not."

The Contrapositive Method

The contrapositive method is a special type of proof by contradiction wherein you assume that A and NOT B are true, and your objective is to reach the contradiction that A is not true. Specifically, you should

1. Write down the statements NOT B and NOT A.
2. Work forward from NOT B to obtain the statement NOT A.

Note: You can work backward from NOT A.

Since the contrapositive method is a type of proof by contradiction, you should use the contrapositive method (even before contradiction) when B contains the key word "no" or "not."

Summary

The Construction Method

The construction method should be used when, in the backward process, you come across the key words "there is" ("there are," "there exists") in the standard form:

> There is an object with a certain property such that something happens.

To use the construction method you should

1. Turn to the forward process to construct, guess, or produce the desired object.
2. Prove that the object you created does satisfy the certain property and the something that happens.

The Uniqueness Methods

You should use a uniqueness method when the conclusion you are trying to establish contains the key word "unique" (or equivalent words), as well as the quantifier "there is" in the standard form:

> There is a unique object with a certain property such that something happens.

The Direct Uniqueness Method

To use this method, you should

1. Use the construction (or contradiction) method to establish the existence of the desired object.
2. Assume that both X and Y are two objects that satisfy the certain property and the something that happens.
3. Work forward from (2) and the hypothesis to prove that X = Y.

Note: You can also work backward from X = Y.

Summary

The Indirect Uniqueness Method

To use this method you should

1. Use the construction (or contradiction) method to establish the existence of the desired object.
2. Assume that X and Y are two different objects that satisfy the certain property and the something that happens.
3. Work forward from (2) and the hypothesis to reach a contradiction.

The proof techniques used in geometry are now complete. You have learned the three major proof techniques for proving propositions, theorems, lemmas, and corollaries. They are the forward-backward, contrapositive, and contradiction methods. You have come to know the quantifier "there is" and the corresponding construction method. For the key word "unique," your bag of proof techniques includes the two uniqueness methods. You are now ready to "speak" geometry. If all of those proof techniques fail, you may wish to stick to Greek -- after all, it's all Greek to me.

Summary Table

Proof Technique	When to Use It	What to Assume	What to Conclude	How to Do It
Forward-Backward	As a first attempt, or when B does not have recognizable key words.	A	B	Work forward from A and apply the backward process to B.
Contra-positive	When B has the key word "no" or "not" in it.	NOT B	NOT A	Work forward from NOT B and backward from NOT A.
Contra-diction	When B has the key word "no" or "not" in it.	A and NOT B	Some contra-diction.	Work forward from A and NOT B to reach a contradiction.
Construc-tion	When B has the key words "there is," "there are," etc.	A	There is the desired object.	Guess, construct, etc. the object having the certain property, and show that the some-thing happens.
Uniqueness (Direct)	When B has the key word "unique" in it.	X and Y are two such objects, and A.	The two objects are equal.	Work forward from A and the properties of X and Y. Also work backward to show that X = Y.
Uniqueness (Indirect)	When B has the key word "unique" in it.	X and Y are two dif-ferent objects, and A.	Some contra-diction.	Work forward from A, the properties of X and Y, and the fact that X and Y are different, to reach a contradiction.

EXERCISES FOR SECTION 8.1

For each of the following problems, indicate which proof technique you would use to begin the proof and briefly explain why.

1. In a plane, there is one and only one line perpendicular to a given line L through a point P on the line.

____

____

2. If p and q are odd integers, then the equation $x^2 + 2px + 2q = 0$ has no rational solution for x.

____

____

3. If two angles of a triangle are equal, then the triangle is isosceles.

____

____

4. If the diagonals of a quadrilateral do not bisect each other, then the quadrilateral cannot be a square.

____

____

5. If x and y are two different real numbers with $x < y$, then there are rational numbers r and s with $x < r < s < y$ such that $|r - s| < 0.001$.

____

____

6. If a line bisects an angle of an equilateral triangle, then the line is perpendicular to the side of the triangle opposite the angle.

____

____

8.2 READING PROOFS

In this book, whenever you had difficulty reading the condensed proof, you could always refer back to the analysis of the proof to understand the steps. In general, reading mathematical proofs in texts can be difficult because

1. A detailed analysis of the proof is not provided.
2. Little or no reference is made to the proof techniques that are being used.
3. Several steps are often condensed into a single sentence with little or no justification.

The next example presents such a condensed proof and shows you how to go about reading and understanding what is happening. The example deals with the concept of congruent triangles.

Definition 16: Two triangles are congruent if and only if all of their corresponding sides and angles are equal.

Note: To determine the "corresponding" sides and angles, you must match up the notation of the two triangles so that each side of one triangle is associated to a specific side of the other triangle.

In the following proof, use will be made of the

ASA Postulate: If two angles and the included side of one triangle are equal to the corresponding angles and side of a second triangle, then the two triangles are congruent.

Example 14: If the line RV intersects TU so as to bisect TU, and UV > RT, then UV is not parallel to RT.

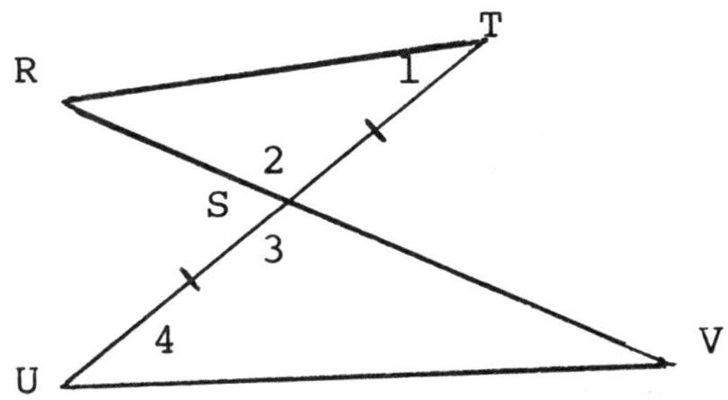

Condensed Proof: (For reference purposes, each sentence of the proof will be written on a separate line.)

S1: Assume that UV is parallel to RT.

S2: It would then follow that $\angle 1 = \angle 4$.

S3: Also, by hypothesis, TS = SU.

S4: Finally, $\angle 2 = \angle 3$, and so triangle RST would be congruent to triangle SUV.

S5: Consequently, UV = RT, which is a contradiction.//

Analysis: Each sentence of the condensed proof will be discussed and explained.

Interpretation of S1

This sentence indicates that either the contrapositive or contradiction method is being used because the author is assuming NOT B. By looking at S5, you can see that the contradiction method is the one that is being used. Can you state precisely what contradiction has been reached?

Summary

The use of the contradicition method should come as no surprise since the conclusion of the proposition contains the key word "not." The remaining steps of the proof should consist of working forward from A and NOT B to reach a contradiction.

Interpretation of S2

It is being claimed that, as a result of assuming that UV is parallel to RT, $\angle 1 = \angle 4$. The author has

1. Failed to give a reason to justify the fact that $\angle 1 = \angle 4$ -- namely, that alternate interior angles of parallel lines cut by a transversal are equal.
2. Failed to state why it is being shown that $\angle 1 = \angle 4$. The reason will become evident in S4.

Interpretation of S3

The author is working forward from A to claim that TS = SU. Again, the author has

1. Failed to justify why TS = SU -- namely, because RV bisects TU, and by the definition of bisection, TS = SU.
2. Failed to state why it is being shown that TS = SU. The reason will become evident in S4.

Interpretation of S4

First, the author is claiming, without justification, that $\angle 2 = \angle 3$. The reason is that vertical angles of intersecting lines are equal.

Then, the claim is being made that the two triangles are congruent. Once again, the author has

Summary

1. Failed to state why the triangles are congruent -- namely, the ASA postulate is being used. Indeed, $\angle 1 = \angle 4$, TS = SU, and $\angle 2 = \angle 3$, as was shown in sentences S2, S3, and S4, respectively. Note that, at the time, no reason was given as to why the author was showing that these angles and sides are equal.
2. Failed to state why it is being shown that the two triangles are congruent. The reason is related to reaching a contradiction, as will be seen in S5.

Interpretation of S5

It is here, at last, that the contradiction is being reached. The first difficulty, however, is trying to understand precisely what the contradiction is. Well, in S5 it says that UV = RT, but in the hypothesis it states that UV > RT, and this indeed is a contradiction.

The only remaining detail of S5 is to figure out the reason that UV = RT. Indeed, UV = RT because the two triangles have been shown to be congruent (see S4), and so, from the definition of congruent triangles, all corresponding sides (including UV and RT) must be equal. Observe that, in S4, no reason was given as to why the two triangles were shown to be congruent.

Example 14 is typical of the way in which mathematical proofs are written. In order to understand them, you must

1. Learn to identify which techniques are being used and how they are applied to the particular problem.
2. Be able to fill in reasons and steps when they are omitted.

When you are having difficulties reading a proof, ask yourself how you would do the proof. Then see if the author is following that same line of reasoning.

8.3 CREATING PROOFS

In this section you will see how the proof techniques of the previous chapters are combined and used in doing proofs. As you read the analysis of the proof, pay particular attention to how the form of the statement under consideration leads to the correct technique.

Once again, the example deals with the concept of congruent triangles (see Definition 16 of the previous section). This time, however, use will be made of the following

SAS Postulate: If two sides and the included angle of one triangle are equal to the corresponding sides and angle of a second triangle, then the two triangles are congruent.

Example 15: If SU is a perpendicular bisector of RT, then triangle RST is isosceles.

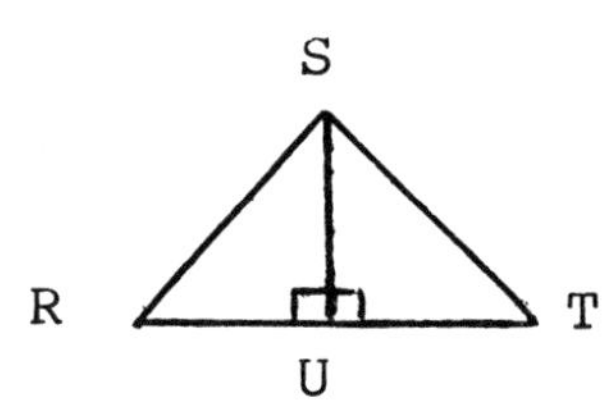

Analysis: Since there are no key words in the conclusion of the proposition, the forward-backward method will be used to begin the proof.

Summary

The Forward-Backward Method

The Backward Process

The conclusion B gives rise to the

Key Question: How can I show that a triangle is isosceles?

By using the definition, one obtains the

Key Answer: Show that at least two sides of the triangle have equal length.

Rewording the key answer and applying it to the specific problem results in the new statement to be proved:

B1: There are at least two sides of triangle RST that have equal length.

Recognizing the appearance of the key words "there are," the construction method will be used.

The Construction Method

The idea now is to turn to the forward process to produce the two sides of triangle RST that have equal length.

The Forward Process

Looking at the figure, it is not hard to guess that the desired sides of the triangle are RS and ST. Now you should use the backward process to show that the something happens, that is, that

B2: RS = ST.

The Backward Process

The figure also indicates an approach for establishing that B2 is true. Specifically, the SAS postulate will be used to show that

B3: Triangles RSU and SUT are congruent,

for then, by definition, all of their corresponding sides (including RS and ST) will be equal. Now, the forward process will be used to establish the congruence of the triangles via the SAS postulate. Can you see which sides and angles are equal?

The Forward Process

By the hypothesis that SU bisects RT, it follows, by definition of a bisector, that

A1: RU = UT.

Also, from the hypothesis that SU is perpendicular to RT, and the definition of perpendicularity, it must be that

A2: $\angle$ RUS and $\angle$ TUS are right angles.

By the definition of right angles,

A3: $\angle$ RUS = 90° = $\angle$ TUS.

Finally, it should be evident that

A4: SU = SU.

Consequently, from A1, A3, A4, and the SAS postulate, triangles RSU and SUT are congruent, which was the last statement of the backward process, thus completing the proof.

Condensed Proof: To prove that triangle RST is isosceles, it will be shown that RS = ST. To that end, the hypothesis that SU is a perpendicular bisector of RT ensures that RU = UT and $\angle$ RUS = $\angle$ TUS. Since SU = SU, it follows that triangles RSU and SUT are congruent, and consequently, RS = ST.//

As a general rule, when the desired conclusion under consideration contains a recognizable key word, it is best to try the associated proof technique. If no key words are apparent, then the forward-backward method is likely to be successful. If that method fails, then you might try asking yourself why the conclusion cannot be false, thus leading you into the contradiction (or contrapositive) method. There is no substitute for creativity, however, the proof techniques are designed to get you started and to guide you along the way. Remember that, as you proceed through a proof, different techniques will be needed as the form of the statement under consideration changes.

SOLUTIONS

CHAPTER 1

1.1 (1) true, false (2) convincing argument (3) another student in the class (4) proof techniques (5) key words

(6), (8), (10), and (11) are valid mathematical statements.

1.2 (1) Hypothesis: n is an even integer.

Conclusion: n^2 is an even integer.

(2) Hypothesis: $\angle 1$ and $\angle 2$ are right angles.

Conclusion: $\angle 1$ and $\angle 2$ are supplementary.

(3) Hypothesis: n is a positive integer.

Conclusion: the sum of the first n integers is $n(n + 1)/2$.

(4) Assume: n is an integer and n^2 is even.

Conclude: n is an even integer.

(5) Assume: p and q are positive real numbers with $\sqrt{pq} \neq (p + q)/2$.

Conclude: $p \neq q$.

(6) Assume: x is a real number.

Conclude: the minimum value of $x(x - 1)$ is at least $-(1/4)$.

(7) Given: x is a real number with $x > 1$.

To Prove: $x > 1/x$.

(8) Given: two integers whose sum is odd.

To Prove: the two integers cannot both be even.

(9) Given: x is a real number with $x \neq 0$.

To Prove: $|x| > 0$.

(10) (d)

1.3 (1) True because $2 > 7$ is false (so $1 > 3$ does not matter).

(2) True because $2 < 7$ is true and $1 < 3$ is true.

(3) True because $1 < 2$ is true (so $x = 3$ does not matter).

(4) True if $x \neq 3$ (because the hypothesis would be false). False if $x = 3$ (because the hypothesis would be true while the conclusion is false).

(5)

A	NOT A
T	F
F	T

(6)

A	B	C	B => C	A => (B => C)
T	T	T	T	T
T	T	F	F	F
T	F	T	T	T
T	F	F	T	T
F	T	T	T	T
F	T	F	F	T
F	F	T	T	T
F	F	F	T	T

(7)

A	B	C	A => B	(A => B) => C
T	T	T	T	T
T	T	F	T	F
T	F	T	F	T
T	F	F	F	T
F	T	T	T	T
F	T	F	T	F
F	F	T	T	T
F	F	F	T	F

(8) You want to show that A is false. Otherwise, if A is true and B is false, then "A implies B" would be false.

CHAPTER 2

2.1 (1) hypothesis (A).

(2) conclusion (B), conclusion (B).

(3) Conclusion (or the last statement in the backward process).

(4) Hypothesis (or the last statement in the forward process).

(5) Turn to the backward process in an attempt to reach the last statement in the forward process.

(6) Turn to the forward process in an attempt to reach the last statement in the backward process.

2.2 (1) ts, st (2) ts (3) 3, ts, 4

(4) ss: $\frac{\text{ssss}}{\text{rule 1}}$ $\frac{\text{NA}}{\text{rule 2}}$ $\frac{\text{NA}}{\text{rule 3}}$ $\frac{\text{sst}}{\text{rule 4}}$

		rule 1	rule 2	rule 3	rule 4
(5)	stt:	sttstt	s	NA	NA
(6)	ttsss:	ttsssttsss	sss	ttt	ttssst
(7)	t:	NA	ttt	sss	NA
(8)	tt:	t	tttt	ssst or tsss	NA
(9)	st:	NA	ttst or sttt	ssss	s

(10) 1, ssss, 3 (11) 2, s, 4 (12) 2 (or 3), sss (or ttt), 3 (or 2)

(13)

s	given
ss	rule 1
ssss	rule 1
sssst	rule 4
tst	rule 3

(14)

tst	given
tsst	rule 1 then rule 2
tssttsst	rule 1
ttst	rule 2 then rule 3

2.3 (1) Key Question: How can I show that an angle is a right angle?

(2) Key Question: How can I show that a triangle is isosceles?

(3) Key Question: How can I show that two lines are perpendicular?

(4) Key Question: How can I show that two angles are complementary?

(5) Key Question: How can I show that two sets are equal?

(6) Key Answer: Show that the degrees of the two angles add up to 180°.
B1: $\angle 1 + \angle 2 = 180°$.

(7) Key Answer: Show that all three sides of the triangle have the same length.
B1: $XY = YZ = XZ$.

(8) Key Answer: Show that the angles formed at the intersection of the two lines are right angles.
B1: $\angle RXT$, $\angle RXU$, $\angle TXS$, and $\angle SXU$ are right angles.

(9) A1: $(x - 1)(x - 2) \leq 0$ A1: $x^2 \leq 3x - 2$ A1: $x(x - 3) + 2 \leq 0$

(10) A1: XZ = ZY A1: $\angle Z = 90°$ A1: $XZ \perp ZY$

(11) A1: TW = WU A1: RW = WS A1: $\angle TWR = 90°$

(12) (c); Uses specific notation from the problem

(13) (a) Correct.
(b) Uses specific notation from the problem.
(c) Uses specific notation from the problem.
(d) An incorrect question for this problem.

2.4 (1) Key Question: How can I show that an integer is even?

Key Question: How can I show that an integer times itself is even?

(2) Key Question: How can I show that two sides of a triangle are equal?

Key Question: How can I show that two real numbers are equal?

(3) Key Question: How can I show that two integers are equal?

Key Question: How can I show that an integer satisfies a quadratic equation?

(4) Key Question: How can I show that a set is a subset of the union of two sets?

Key Question: How can I show that a set is a subset of another set?

(5) Key Answer: Show that the ratio of the two numbers is 1.

Key Answer: Show that the first number is less than or equal to the second number, and the second number is less than or equal to the first number.

(6) Key Answer: Show that the triangle has a right angle.

Key Answer: Show that two sides of the triangle are perpendicular.

(7) Key Answer: Show that they have the same number of degrees.

Key Answer: Show that they are both equal to a third angle.

(8) Key Question: How can I show that a triangle is equilateral?

Key Answer: Show that all three sides of the triangle have the same length.
B1: RS = ST = RT

(9) Key Question: How can I show that two angles are supplementary?

Key Answer: Show that the degrees of the two angles add up to 180°.
B1: $\angle USV + \angle RST = 180°$.

(10) Key Question: How can I show that the sum of the degrees of two angles is 90°.

Key Answer: Show that the sum of the degrees of the two angles plus 90° equals 180°.

(11) Key Question: How can I show that the difference of two real numbers is 0?

Key Answer: Show that the square of the difference of the two real numbers is 0.

2.5 (1) Key Question: How can I show that two lines are parallel?

Key Question: How can I show that two lines tangent to a circle at the endpoints of a diameter are parallel?

(2) Key Question: How can I show that a function is polynomial?

Key Question: How can I show that the sum of two functions is polynomial?

(3) Key Answer: Show that angle-side-angle.

Key Answer: Show that side-angle-side.

Key Answer: Show that side-side-side.

(4) Key Answer: Show that they lie in the same plane and do not intersect.

Key Answer: Show that they are both parallel to a third line.

Key Answer: Show that they are both perpendicular to a third line.

(5) Key Answer: Show that the quadrilateral has four right angles.

Key Answer: Show that the quadrilateral is a parallelogram with one right angle.

Key Answer: Show that the adjacent sides of the quadrilateral are perpendicular.

(6) Key Question: How can I show that a solution to a quadratic equation is positive?

Key Answer: Show that a root obtained by the quadratic formula is positive.
B1: $(-b \pm \sqrt{b^2 - 4ac})/(2a) > 0$.

(7) Key Question: How can I show that a triangle is equilateral?

Key Answer: Show that all three sides of the triangle have the same length.
B1: RS = ST = RT.

(8) A1: Center = (3, 2) A1: Radius = 5

A1: $x^2 - 6x + 9 + y^2 - 4y + 4 = 25$

(9) A1: $\cos(X) = 1/\sqrt{2}$ A1: $\angle X = 45°$ A1: $y = 1$

(10) A1: UV = VW A1: $\angle U = 60°$ A1: Area = $\sqrt{3}$ UV/4

(11) (d); Explain your choice. It is not stated that $x \neq 5$, so, if $x = 5$, you will not be able to divide by $x - 5$.

2.6 (1) Condensed Proof: Since RS is a line, $\angle 2 + \angle 3 = 180°$. Also, it is given that $\angle 3 = \angle 4$ so $\angle 2 + \angle 4 = 180°$. Furthermore, since $\angle 4$ and $\angle 1$ are vertical angles of intersecting lines, $\angle 4 = \angle 1$. Thus, $\angle 2 + \angle 1 = 180°$, and so $\angle 2$ and $\angle 1$ are supplementary by definition.

(2) Condensed Proof: From the hypothesis, $\angle 1$ and $\angle 2$ are complementary so, by definition, $\angle 1 + \angle 2 = 90°$. For similar reasons, $\angle 3 + \angle 4 = 90°$, thus, $\angle 1 + \angle 2 = \angle 3 + \angle 4$. Finally, it is given that $\angle 2 = \angle 4$ so, by subtracting, the desired conclusion that $\angle 1 = \angle 3$ is obtained.

(3) Given: $\angle TSU = \angle USV$ and RU is a line.

To Prove: $\angle USV$ and $\angle RST$ are supplementary.

PROOF

STATEMENT	REASON
1. RU is a line.	1. Given.
2. $\angle RST + \angle TSU = 180°$.	2. $\angle RST + \angle TSU = \angle RU$, and RU is a line.
3. $\angle TSU = \angle USV$.	3. Given.
4. $\angle RST + \angle USV = 180°$.	4. Substitute (3) into (2).
5. $\angle RST$ and $\angle USV$ are supplementary.	5. Definition of supplementary angles.

Condensed Proof: Since RU is a line, and $\angle RST + \angle TSU = \angle RU$, it follows that $\angle RST + \angle TSU = 180°$. Given that $\angle TSU = \angle USV$, one has $\angle RST + \angle USV = 180°$. Thus, by the definition of supplementary angles, $\angle RST$ and $\angle USV$ are supplementary.

(4) Given: $RS \perp ST$ To Prove: $\angle 3 = \angle 4$

$ST \perp TU$

$\angle 1 = \angle 2$

PROOF

STATEMENT	REASON
1. $RS \perp ST$.	1. Given.
2. $\angle RST = 90°$.	2. Definition of perpendicularity.
3. $\angle 1 + \angle 3 = 90°$.	3. Because $\angle 1 + \angle 3 = \angle RST$.
4. $\angle 1 = \angle 2$.	4. Given.
5. $\angle 2 + \angle 3 = 90°$.	5. Substitute (4) into (3).
6. $ST \perp TU$.	6. Given.
7. $\angle STU = 90°$.	7. Definition of perpendicularity.
8. $\angle 2 + \angle 4 = 90°$.	8. Because $\angle 2 + \angle 4 = \angle STU$.
9. $\angle 2 + \angle 4 = \angle 2 + \angle 3$.	9. From (5) and (8).
10. $\angle 4 = \angle 3$.	10. Subtract $\angle 2$ from both sides of (9).

(5) Given: R and S are two intersecting lines.

To Prove: $\angle 1 = \angle 3$ and $\angle 2 = \angle 4$.

PROOF

STATEMENT	REASON
1. S is a line.	1. Given.
2. $\angle 1 + \angle 2 = 180°$.	2. A line has 180°.
3. R is a line.	3. Given.
4. $\angle 2 + \angle 3 = 180°$.	4. A line has 180°.
5. $\angle 1 + \angle 2 = \angle 2 + \angle 3$.	5. From (2) and (4).
6. $\angle 1 = \angle 3$.	6. Subtract $\angle 2$ from both sides of (5).
7. $\angle 3 + \angle 4 = 180°$.	7. A line has 180°.
8. $\angle 3 + \angle 4 = \angle 2 + \angle 3$.	8. From (7) and (4).
9. $\angle 2 = \angle 4$.	9. Subtract $\angle 3$ from both sides of (9).

Condensed Proof: Since R and S are lines, it follows that $\angle 1 + \angle 2 = \angle 2 + \angle 3 = 180°$. By subtracting $\angle 2$ from both sides, one has that $\angle 1 = \angle 3$. Similarly, since R and S are lines, $\angle 2 + \angle 3 = \angle 3 + \angle 4 = 180°$. By subtracting $\angle 3$ from both sides, one has that $\angle 2 = \angle 4$.

(6) Given: $RS \perp RU$.
$ST \perp TU$.
$\angle 1 = \angle 2$.

To Prove: $\angle 3 = \angle 4$.

PROOF

STATEMENT	REASON
1. $RS \perp RU$.	1. Given.
2. $\angle 1 + \angle 3 = \angle URS = 90°$.	2. Definition of perpendicularity.
3. $\angle 1 = \angle 2$.	3. Given.
4. $\angle 2 + \angle 3 = 90°$.	4. Substitute (3) into (2).
5. $ST \perp TU$.	5. Given.
6. $\angle 2 + \angle 4 = \angle UTS = 90°$.	6. Definition of perpendicularity.
7. $\angle 2 + \angle 3 = \angle 2 + \angle 4$.	7. From (4) and (6), both equal 90°.
8. $\angle 3 = \angle 4$.	8. Subtract $\angle 2$ from both sides of (7).

Condensed Proof: By hypothesis, RS is perpendicular to RU. Thus, from the definition of perpendicularity, $\angle 1 + \angle 3 = 90°$. Also, it is given that $\angle 1 = \angle 2$, so by substituting, $\angle 2 + \angle 3 = 90°$. Similarly, by hypothesis, ST is perpendicular to TU, and so $\angle 2 + \angle 4 = 90°$. But then, $\angle 2 + \angle 3 = \angle 2 + \angle 4$ and so $\angle 3 = \angle 4$.

(7) Analysis: The forward-backward method gives rise to the
Key Question: How can I show that a triangle (namely RST) is equilateral?

One answer is to show that all three sides have equal length, so the new statement to be proved is

B1: RS = ST = RT.

To see that RS = ST, it will be shown that

B2: RS = 2SU and ST = 2SU.

The fact that RS = 2SU is given in the hypothesis. Also from the hypothesis that RU is a bisector of ST, one has

A1: SU = UT.

Consequently,

A2: ST = SU + UT = SU + SU = 2SU.

To obtain B1, it remains to show that

B3: RS = RT.

To do so, it will be shown by the side-angle-side postulate that

B4: Triangle RSU is congruent to triangle RUT.

From A1 one has SU = UT. Also, from the hypothesis, RU is perpendicular to ST, and so

A3: $\angle RUS = \angle RUT = 90°$.

Finally, RU = RU, so indeed the triangles are congruent, and so the proof is complete.

Condensed Proof: To reach the conclusion, it will be shown that RS = ST = RT. Since RU is a bisector of ST, SU = UT. Thus ST = SU + UT = SU + SU = 2SU. Also, it is given that RS = 2SU, so RS = ST. To see that RS = ST, note that triangle RSU is congruent to RUT. This is so because SU = UT (as noted above). Since RU is perpendicular to ST, $\angle RUS = \angle RUT = 90°$, and also, RU = RU. By the side-angle-side postulate, the two triangles are congruent and so RS = ST.

(8) Analysis: The forward-backward method gives rise to the
Key Question: How can I show that the area of a triangle (namely XYZ) equals a number (namely $z^2/4$)?

One answer is to use the formula that the area of a triangle is one-half the base times the height. Thus, the new statement to be proved is

B1: $xy/2 = z^2/4$.

Working forward from the hypothesis that the triangle is isosceles, it follows by definition that

A1: $x = y$.

Consequently, the area of triangle XYZ is

A2: Area = $xy/2 = x^2/2$.

The Pythagorean theorem states that $x^2 + y^2 = z^2$ so, from A1,

A3: $x^2 + x^2 = z^2$, or, $2x^2 = z^2$, or $x^2 = z^2/2$.

Replacing x^2 in A2 by $z^2/2$ from A3, one obtains precisely B1, thus completing the proof.

Condensed Proof: From the formula for the area of a triangle, the area of XYZ = $xy/2$. But since the triangle is isosceles, $x = y$, so its area is $x^2/2$. The Pythagorean theorem yields $x^2 + y^2 = z^2$ but, since $x = y$, one has $2x^2 = z^2$, or $x^2 = z^2/2$. Thus area of XYZ = $x^2/2 = z^2/4$.

2.7 (1) Forward, backward, forward.

(2) Backward, backward, forward, forward.

(3) Forward, backward, forward.

(4) PROOF

STATEMENT	REASON
1. $\angle 1$ and $\angle 2$ are complementary.	1. Given.
2. $\angle 1 + \angle 2 = 90°$.	2. Definition of complementary angles.
3. $\angle 3$ and $\angle 4$ are complementary.	3. Given.
4. $\angle 3 + \angle 4 = 90°$.	4. Definition of complementary angles.
5. $\angle 1 + \angle 2 = \angle 3 + \angle 4$.	5. From (2) and (4).
6. $\angle 2 = \angle 4$.	6. Given.
7. $\angle 1 = \angle 3$.	7. Subtract (6) from (5).

(5) PROOF

STATEMENT	REASON
1. $RS \perp ST$.	1. Given.
2. $\angle RST = \angle 2 + \angle 3 = 90°$.	2. Definition of perpendicularity.
3. $\angle 1 = \angle 3$.	3. Given.
4. $\angle 2 + \angle 1 = 90° = \angle XSY$.	4. Substitute (3) into (2).
5. $XS \perp SY$.	5. Definition of perpendicularity.

(6) PROOF

STATEMENT	REASON
1. RS is a line.	1. Given.
2. $\angle 1 + \angle 2 = \angle RS = 180°$.	2. A line has 180°.
3. TU is a line.	3. Given.
4. $\angle 2 + \angle 3 = \angle TU = 180°$.	4. A line has 180°.
5. $\angle 1 + \angle 2 = \angle 2 + \angle 3$.	5. From (2) and (4).
6. $\angle 1 = \angle 3$.	6. Subtract $\angle 2$ from both sides of (5).

(7) PROOF

STATEMENT	REASON
1. RS is a line.	1. Given.
2. $\angle 3 + \angle 4 = \angle RS = 180°$.	2. A line has 180°.
3. $\angle 1 = \angle 2$.	3. Vertical angles of intersecting lines are equal.
4. $\angle 2 = \angle 3$.	4. Given.
5. $\angle 1 = \angle 3$.	5. From (3) and (4).
6. $\angle 1 + \angle 4 = 180°$.	6. Substitute (5) into (2).
7. $\angle 1$ and $\angle 4$ are supplementary.	7. Definition of supplementary angles.

3.1 (1) Unacceptable - uses the term being defined within the definition.

(2) Acceptable.

(3) Unacceptable - the term "evenly" is not an acceptable undefined term.

(4) Unacceptable - does not contain enough information to distinguish a median.

(5) Unacceptable - does not contain the least amount of information.

(6) It can be expressed as two times some integer.

(7) It can be expressed as two times some integer, plus one.

(8) It can be expressed as the ratio of two integers in which the denominator is not 0.

(9) A quadrilateral is a polygon with four sides.

(10) A parallelogram is a quadrilateral in which opposite sides are parallel to each other.

(11) A rectangle is a parallelogram with one right angle.

(12) A convex polygon is a polygon with the property that between any two points in the polygon, the line segment joining those points lies within the polygon.

CHAPTER 3

3.2 (1) Key Question: How can I show that an integer is odd?

Key Answer: Show that the integer can be expressed as two times some integer, plus one.

B1: $mn = 2k + 1$ for some integer k.

(2) Key Question: How can I show that a triangle is isosceles?

Key Answer: Show that at least two of its sides have equal length.

B1: $XZ = YZ$.

(3) Key Question: How can I show that two lines are parallel?

Key Answer: Show that they lie in the same plane and that they do not intersect.

B1: R and S lie in the same plane, and R and S do not intersect.

(4) Key Question: How can I show that a number is a rational number?

Key Answer: Show that the number can be expressed as the ratio of two integers in which the denominator is not 0.

B1: $s + t = p/q$ where p and q are integers and $q \neq 0$.

(5) A1: $n = 2k + 1$ for some integer k.

(6) A1: $XY = YZ = XZ$.

(7) A1: $p = a/b$ where a and b are integers and $b \neq 0$.

(8) A1: $n^2 = 2k$ for some integer k.

(9) Key Question: How can I show that an integer (namely mn) is odd?

B1: $mn = 2k + 1$ for some integer k.

A1: $m = 2p + 1$ for some integer p.

A2: $n = 2q + 1$ for some integer q.

A3: $mn = (2p + 1)(2q + 1)$.

A4: $mn = 4pq + 2p + 2q + 1 = 2(2pq + p + q) + 1$.

From A4, it follows that the desired value of the integer k is $2pq + p + q$, thus completing the proof.

PROOF

STATEMENT	REASON
1. m and n are odd integers.	1. Given.
2. $m = 2p + 1$ for some integer p.	2. Definition of an odd integer.
3. $n = 2q + 1$ for some integer q.	3. Definition of an odd integer.
4. $mn = (2p + 1)(2q + 1)$.	4. Multiplying (2) by (3).
5. $mn = 2(2pq + p + q) + 1$.	5. Rewriting (4).
6. $mn = 2k + 1$ where $k = 2pq + p + q$, and so mn is odd.	6. Definition of an odd integer.

(9) cont'd.

mn = 2k + 1, m = 2p + 1, n = 2q + 1.

mn = (2p + 1)(2q + 1) = 4pq + 2p + 2q + 1 = 2(2pq + p + q) + 1.

Therefore, the desired value for k is 2pq + p + q, and mn is odd.

(10) PROOF

STATEMENT	REASON
1. Triangle RST is equilateral.	1. Given.
2. RS = ST = RT.	2. Definition of an equilateral triangle.
3. RS = RU.	3. Given.
4. ST = TU.	4. Given.
5. RS = TU.	5. From (2) and (3).
6. RT = RU = TU.	6. From (3) and (5).
7. Triangle RUT is equilateral.	7. Definition of an equilateral triangle.

Condensed Proof: By hypothesis, triangle RST is equilateral. Consequently, RS = ST = RT. Also, it is given that RS = RU and ST = TU. Hence, RU = TU = RT and so, by definition, triangle RUT is equilateral.

3.3 (1) Reversible.

(2) Not reversible.

(3) Reversible.

(4) Not reversible.

(5) Definition 9: If r can be expressed as the ratio of two integers p and q in which the denominator q is not 0, then R is a rational number.

(6) Definition 10: If two lines lie in the same plane and they do not intersect, then the two lines are parallel.

(7) Statement 1: If all three sides of a triangle are equal, then all three angles of the triangle are equal.

Statement 2: If all three angles of a triangle are equal, then all three sides of the triangle are equal.

(8) Statement 1: If two lines lie in the same plane and do not intersect, then the two lines lie in the same plane and have the same slope.

Statement 2: If two lines lie in the same plane and have the same slope, then they lie in the same plane and they do not intersect.

(9) True; Converse: If a rectangle is a square, then the rectangle has two adjacent sides of equal length; Converse is true.

(10) False, Converse: If $x > 4$, then $|x| > 4$; Converse is true.

(11) True, Converse: If $xy = 0$, then $x = 0$; Converse is false.

(12) False, Converse: If $\angle 1$ and $\angle 2$ are supplementary, then $\angle 1$ and $\angle 2$ are acute; Converse is false.

(13) Assume: n is an even integer.

Conclude: n^2 is an even integer.

(14) Assume: In triangle XYZ, $\angle X = \angle Y$.

Conclude: Triangle XYZ is isosceles.

(15) Analysis: The forward-backward method gives rise to the
Key Question: How can I show that a statement (A) implies another statement (C)?

According to the truth table, the answer is to assume that

A1: A is true.

Your objective is to conclude that

B1: C is true.

Working forward from the hypothesis that "A implies B," and using the truth table and A1, one can say that

A2: B is true.

Similarly, from the hypothesis that "B implies C," and using the truth table and A2, one can conclude that B1 is true, thus completing the proof.

Condensed Proof: To reach the desired conclusion, assume that A is true. From the hypothesis that "A implies B," it then follows that B is true. Similarly, since "B implies C," and B is true, one can conclude that, indeed, C is true.

(16) Analysis: First, it will be shown that A is equivalent to B. The forward-backward method gives rise to the
Key Question: How can I show that two statements (namely A and B) are equivalent?

According to the definition, one must now show that

B1: "A implies B," and "B implies A."

Working forward from the hypothesis, one immediately has that "A implies B," so it remains to show that

B2: "B implies A."

To do so, one can use the hypothesis and the previous exercise to claim that since "B implies C" and "C implies A," it follows that "B implies A" and so B2 is true. In order to complete the proof, it must still be shown that A is equivalent to C. The analysis is similar to the one given above and will not be included here.

Condensed Proof: First it will be shown that A is equivalent to B. From the hypothesis, "A implies B." Also, since "B implies C" and "C implies A, by the previous exercise, it follows that "B implies A," and thus A is equivalent to B. In a similar manner, it can be seen that A is equivalent to C. In the hypothesis it states that "C implies A." Also, since "A implies B" and "B implies C," by the previous exercise, "A implies C."

(17) 6; 4; By using Exercises 15 and 16 it can easily be shown that each statement implies all of the other statements and hence they are all equivalent.

3.4 (1) Example 2.

(2) Example 1.

(3) Example 1.

(4)

Current Notation		Notation of Example 1
E	<---------------------->	R
F	<---------------------->	S
G	<---------------------->	T
H	<---------------------->	U
$\angle EFG$	<---------------------->	$\angle RST$
$\angle EFH$	<---------------------->	$\angle RSU$

$\angle EFG$ and $\angle EFH$ are right angles.

(5)

Current Notation		Notation of Example 2
F	<---------------------->	R
E	<---------------------->	S
G	<---------------------->	T
$\angle EFG$	<---------------------->	$\angle 1$
$\angle GEF$	<---------------------->	$\angle 2$
$\angle EGF$	<---------------------->	$\angle 3$

The line segments EG and FG are perpendicular and $\angle EFG + \angle GEF + \angle EGF = 180°$.

(6)

Current Notation		Notation of Example 3
U	<---------------------->	X
V	<---------------------->	Y
W	<---------------------->	Z
u	<---------------------->	x
v	<---------------------->	y
w	<---------------------->	z

Triangle UVW is a right triangle that has an area of $w^2/4$.

(7) Analysis: The forward-backward method gives rise to the Key Question: How can I show that two angles are supplementary.

One answer is supplied by the conclusion of Example 1, whereby it must by shown that

B1: $\angle EFG$ and $\angle EFH$ are right angles.

Working forward from the hypothesis in Exercise 4, it states that triangle EFH is a right triangle. Consequently,

A1: $\angle EFH$ is a right angle.

Also, by that hypothesis, $\angle GFI$ and $\angle IFE$ are complementary. Hence, by the definition of complementary angles,

A2: $\angle GFI + \angle IFE = \angle EFG = 90°$.

So, by the definition of right angles,

A3: $\angle EFG$ is a right angle.

Since A1 and A3 together are B1, the proof is complete.

Condensed Proof: Since triangle EFH is a right triangle, $\angle EFH$ is a right angle. Also, since $\angle GFI$ and $\angle IFE$ are complementary, it follows that $\angle GFI + \angle IFE = 90°$. Thus, $\angle EFG$ is a right angle. Consequently, the hypothesis (and hence the conclusion) of Example 1 is satisfied, and thus $\angle EFG$ and $\angle EFH$ are supplementary.

(8) Analysis: The forward-backward method gives rise to the Key Question: How can I show that two angles are complementary.

One answer is supplied by the conclusion of Example 2, whereby, it must be shown that

B1: EG and FG are perpendicular, and $\angle EFG + \angle GEF + \angle EGF = 180°$.

Working forward from the hypothesis that triangle EFG is a right triangle it follows that

A1: $\angle EGF$ is a right angle.

From A1 and the definition of perpendicularity, one has

A2: EG and FG are perpendicular.

Finally, the hypothesis of Exercise 5 also states that

A3: $\angle EFG + \angle GEF + \angle FGE = 180°$.

Since A2 and A3 are B1, it follows that the hypothesis (and hence the conclusion) of Example 2 is true. Thus, $\angle EFG$ and $\angle GEF$ are complementary.

Condensed Proof: Since triangle EFG is a right triangle, $\angle EGF$ is a right angle, and so EG is perpendicular to FG. Also, it states in the hypothesis that $\angle EFG + \angle GEF + \angle FGE = 180°$. Consequently, the hypothesis (and hence the conclusion) of Example 2 is satisfied, and thus $\angle EFG$ and $\angle GEF$ are complementary.

(9) Analysis: The forward-backward method gives rise to the Key Question: How can I show that a triangle is isosceles?

One answer is by the conclusion of Example 3, whereby it must be shown that

B1: The right triangle UVW has an area of $w^2/4$. Working forward from the hypothesis in Exercise 6, since $\sin(U) = \sqrt{u/2v}$, it follows that

A1: $u/w = \sqrt{u/2v}$, or

A2: $u^2/w^2 = u/2v$, or

A3: $uv = w^2/2$.

So, the area of UVW is one-half the base times the height, and consequently, from A3,

A4: area of UVW $= uv/2 = w^2/4$.

Thus, the proof is complete.

Condensed Proof: Since $\sin(U) = \sqrt{u/2v}$, it follows that $u/w = \sqrt{u/2v}$, or $u^2/w^2 = u/2v$, or $uv = w^2/2$. Also, the area of the right triangle UVW $= uv/2 = w^2/4$. Hence the hypothesis (and hence the conclusion) of Example 3 is satisfied, and thus triangle UVW is isosceles.

CHAPTER 4

4.1 (1) NOT A: n^2 is an odd integer.

(2) NOT A: $x \neq y$.

(3) NOT A: The set S has at least six elements.

(4) NOT A: Triangle RST is not isosceles and it is not equilateral.

(5) NOT A: Triangle RST is either isosceles or equilateral.

(6) NOT A: Either $\angle X = \angle Y$ or $\angle X$ is not acute.

(7) NOT A: r and s are rational numbers but r + s is not a rational number.

(8) $\angle R$ is not acute and it is not a right angle.

(9) n is not an even integer.

(10) No two sides of triangle XYZ have different length.

(11) x is not less than y and x is not greater than y.

(12) It is not true that $x \geq 5$ and $x \leq -5$.

(13) It is not true that $x \leq 5$ and $x \geq -5$.

(14) There are not a finite number of even integers.

4.2 (1) Assume: R is perpendicular to T, S is perpendicular to T, and R and S do intersect.

(2) Assume: n is an integer and n^2 is odd, and n is even.

(3) Assume: $\angle R \neq \angle S$, and $\angle R$ and $\angle S$ are vertical angles.

(4) Assume: r is a real number such that $r^2 = 2$ and r is a rational number.

(5) Assume: x and y are real numbers such that $x \geq 0$, $y \geq 0$, $x + y = 0$, and either $x \neq 0$ or $y \neq 0$.

(6) Condensed Proof: Assume that an integer n is both odd and even. Then, by definition, $n = 2k + 1$ for some integer k, and $n = 2p$ for some integer p. But then $2p = 2k + 1$, or $2(p - k) = 1$, or $p - k = 1/2$. But it is impossible for the difference of two integers to be 1/2.

(7) Condensed Proof: Assume that x, $y > 0$ and $x + y = \sqrt{x^2 + y^2}$. On squaring both sides, $x^2 + 2xy + y^2 = x^2 + y^2$, or $2xy = 0$. But since x, $y > 0$, $2xy > 0$ and so, the contradiction is that $0 = 2xy > 0$.

(8) Condensed Proof: Assume that RS is a line, $\angle 1$ and $\angle 3$ are complementary, and $\angle 2$ is obtuse. Then, by the definition of complementary angles, $\angle 1 + \angle 3 = 90°$. Also, since $\angle 2$ is obtuse, $\angle 2 > 90°$. Thus, $\angle 1 + \angle 2 + \angle 3 > 180°$, but because $RS = \angle 1 + \angle 2 + \angle 3$ is a line, it must be that $\angle 1 + \angle 2 + \angle 3 = 180°$.

(9) Condensed Proof: Assume that x, $y \geq 0$ and $x + y = 0$, and either $x \neq 0$ or $y \neq 0$. Suppose that $x \neq 0$. Then $x > 0$, but then $x + y > 0$, and this contradicts the fact that $x + y = 0$. Similarly, if $y \neq 0$, then $y > 0$ and again, $x + y > 0$, a contradiction.

(10) Condensed Proof: Assume that $\angle 1$ and $\angle 2$ are complementary, and that $\angle 3$ is not acute. Then $\angle 3 \geq 90°$. Also, $\angle 2 = \angle 3$ because they are vertical angles. Finally, $\angle 1 + \angle 2 = 90°$ because they are complementary. Thus one would have $90° = \angle 1 + \angle 2 = \angle 1 + \angle 3 \geq \angle 1 + 90°$. By subtracting 90° from both sides, one obtains the contradiction that $\angle 1 \leq 0°$.

4.3 (1) Key Answer: Show that the lines lie in the same plane and that they do not intersect.

Key Answer: Show that the two lines lie in the same plane and they are both perpendicular to a third line.

(2) Key Answer: Show that the integer is not odd.

Key Answer: Show that the integer can be expressed as two times some integer.

(3) Key Answer: Show that the angle is not obtuse and that it is not a right angle.

Key Answer: Show that the angle has less than 90°.

(4) Condensed Proof: Assume that a triangle has at least two obtuse angles, say $\angle 1$ and $\angle 2$. But then, $\angle 1 > 90°$, $\angle 2 > 90°$, and so $\angle 1 + \angle 2 + \angle 3 > 180° + \angle 3 > 180°$, contradicting the fact that the angles of a triangle must add up to 180°.

(5) Condensed Proof: Assume that $\angle 1$ and $\angle 2$ are supplementary, and that R and S are not parallel. Since $\angle 1$ and $\angle 2$ are supplementary $\angle 1 + \angle 2 = 180°$. Since R and S are not parallel, they intersect to form a triangle whose third angle is $\angle 3$. But then, $\angle 1 + \angle 2 + \angle 3 = 180° + \angle 3 > 180°$, contradicting the fact that the angles of a triangle must add up to 180°.

(6) Condensed Proof: Assume that all the angles of triangle RST are equal, but those of triangle TUV are not. Also, assume that RS is parallel to UV. But then $\angle RST = \angle TUV$, $\angle SRT = \angle UVT$, and $\angle RTS = \angle UTV$. Also, since a triangle has 180°, $\angle RST = \angle SRT = \angle RTS = 60°$. Thus, all the angles of triangle TUV would be equal, a contradiction.

4.4 (1) Condensed Proof: Assume that a chord E of a circle is strictly longer than a diameter. Construct a triangle having one side being the chord E, and another side being a diameter D, and the third side, say F. Triangle DEF is an inscribed triangle with diameter D, and so it is a right triangle. But then, E, a leg, would be strictly longer than the hypotenuse D, a contradiction.

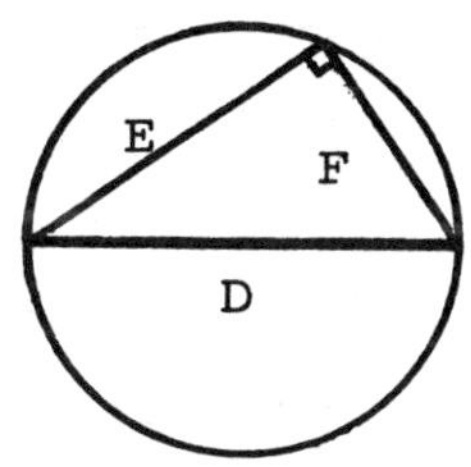

(2) Condensed Proof: Assume that a and b are distinct positive integers such that a divides b. Also assume that b divides a. Then, by definition, $b = ca$ for some integer c, and $a = bd$ for some integer d. But then $b = ca = cbd$, and since $b > 0$, $1 = cd$. This means that $c = 1$ and $d = 1$. But then, $b = a$, contradicting the assumption that a and b are distinct.

(3) Condensed Proof: Assume that n, n + 1, and n + 2 are three consecutive positive integers such that $(n + 2)^3 = n^3 + 6n^2 + 12n + 8 = n^3 + (n + 1)^3 = n^3 + n^3 + 3n^2 + 3n + 1 = 2n^3 + 3n^2 + 3n + 1$. Then $n^3 - 3n^2 - 9n = 7$, so $n(n^2 - 3n - 9) = 7$, and so either $n = 1$ or $n^2 - 3n - 9 = 1$. Clearly $n = 1$ does not work, so $n^2 - 3n - 9 = 1$. But this quadratic equation has no integer solution for n.

(4) Condensed Proof: Assume that triangles XYZ and VWX are not similar, and that VW is parallel to YZ. Then $\angle VXW = \angle YXZ$, $\angle XVW = \angle XYZ$, and $\angle XWV = \angle XZY$. Hence the two triangles would be similar, a contradiction.

CHAPTER 5

5.1 (1) Object: Mountain.

Certain Property: Over 20,000 feet high.

Something Happens: The mountain is taller than every other mountain in the world.

(2) Object: Integer x.

Certain Property: None.

Something Happens: $2x^2 - 5x + 3 = 0$.

(3) Object: Point.

Certain Property: In the interior of a square.

Something Happens: The point is equally distant from the four corners of the square.

(4) Object: Acute angles.

Certain Property: None.

Something Happens: The acute angles have the same number of degrees.

(5) Object: Distinct rational numbers.

Certain Property: Between x and y.

Something Happens: $|r - s| < 0.001$.

(6) Object: Element x.

Certain Property: In the set of S.

Something Happens: x is not in the set T.

(7) Object: Point S.

Certain Property: In the polygon P.

Something Happens: If R and T are points in P, and S is on the line segment from R to T, then R = S = T.

(8) In an isosceles triangle, there are two angles having the same number of degrees.

(9) A rational number r is a number for which there are integers p and q with $q \neq 0$ such that $r = p/q$.

(10) At a party of n people, there are at least two people who have the same number of friends at the party.

(11) In a triangle, there is an acute angle.

(12) For a polynomial p(x) of degree n, there are exactly n complex roots $r_1, \ldots, r_n$ such that $p(r_1) = \ldots = p(r_n) = 0$.

(13) Unique.

(14) Unique.

(15) Not unique.

(16) Not unique.

(17) Proof Technique: Forward-Backward because B does not contain key words.

(18) Proof Technique: Construction because of the hidden key words "there is."

(19) Proof Technique: Contradiction because B contains the key word "not."

(20) Proof Technique: Forward-Backward because B does not contain key words.

5.2 (1) The value $-3/2$ is not an integer, so the wrong object has been produced.

(2) The value 2 is not less than 0, so this value of x does not satisfy the certain property.

(3) Key Question: How can I show that a number (namely rx) is rational?

Certain Property: $q \neq 0$.

Something Happens: $rx = p/q$.

A1: There are integers s and t with $t \neq 0$ such that $r = s/t$.

A2: There are integers y and z with $z \neq 0$ such that $x = y/z$.

Condensed Proof: Since r and x are rational numbers, by definition, there are integers s, t, y, and z with $t \neq 0$ and $z \neq 0$ such that $r = s/t$ and $x = y/z$. Consequently, on setting $p = sy$ and $q = tz$, and noting that $q \neq 0$, it follows that $p/q = (sy)/(tz) = (s/t)(y/z) = rx$. Thus, rx is rational by definition.

(4) Analysis: The forward-backward method gives rise to the Key Question: How can I show that a number (namely $a + d$) is rational?

Using the definition, it will be shown that

B1: There are integers p and q with $q \neq 0$ such that $a + d = p/q$.

Recognizing the key words "there are" in B1, the construction method will be used to produce p and q. Turning to the forward process, by definition of a and d being rational,

A1: There are integers b and c with $c \neq 0$ such that $a = b/c$.

A2: There are integers e and f with $f \neq 0$ such that $d = e/f$.

By adding the expression for a and d from A1 and A2, one has

A3: $a + d = (b/c) + (e/f) = (bf + ec)/cf$.

Hence, it appears that the desired integers p and q are $bf + ec$ and cf, respectively. To verify the certain property and the something that happens in B1, it must be shown that, for the constructed values of p and q,

B2: $cf \neq 0$.

B3: $a + d = p/q$.

To see that B2 is true note that, from A1 and A2, $c \neq 0$ and $f \neq 0$. Therefore, $cf \neq 0$. Finally, from A1 and A2 one obtains precisely B3 because $p/q = (bf + ce)/cf = (b/c) + (e/f) = a + d$.

Condensed Proof: Since a and d are rational numbers, by definition, there are integers b, c, e, and f with $c \neq 0$, and $f \neq 0$ such that $a = b/c$ and $d = e/f$. Let $p = bf + ec$ and $q = cf$. Noting that $q \neq 0$ because $c \neq 0$ and $f \neq 0$, it also follows that $a + d = (b/c) + (e/f) = (bf + ec)/cf = p/q$. Thus, by definition, $a + d$ is a rational number.

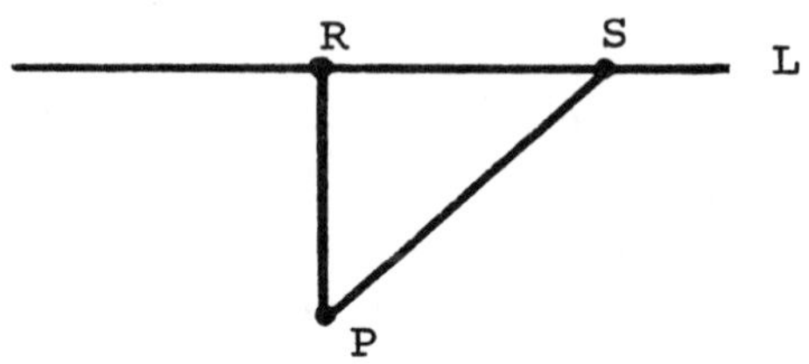

(5) Analysis: Recognizing the key words "there is" in the conclusion, the construction method will be used to produce the desired right triangle. Turning to the forward process, and using the fact that P is not on the line L, there is a postulate stating that

A1: There is a line through P perpendicular to L.

Suppose that the line in A1 intersects L at the point R. Using that line, choose any point S on L other than R, and construct the triangle PRS. To complete the proof, it must be shown that

B1: PRS is a right triangle having P as a vertex and one side on the line L.

From A1, since PR is perpendicular to L, by definition,

A2: $\angle PRS$ is a right angle.

By the definition of a right angle, it follows that

A3: Triangle PRS is a right triangle.

Finally, it is clear that P is a vertex of triangle PRS, and the side RS of triangle PRS lies on the line L. Thus, B1 has been established and the proof is complete.

Condensed Proof: By the appropriate postulate, since P is not on the line L, there is a line through P perpendicular to L meeting L at some point, say R. Choose any point S on L with $S \neq R$. Then the triangle PRS is a right triangle, for $\angle PRS = 90°$. Clearly P is a vertex of triangle PRS, and the side RS lies on the line L.

(6) Analysis: The forward-backward method gives rise to the Key Question: How can I show that an integer (namely a) divides another integer (namely c)?

Using the definition leads to the new statement to be proved:

B1: There is an integer f such that $c = af$.

Recognizing the key words "there is," the construction method will be used. Turning to the forward process, since a divides b, by definition,

A1: There is an integer d such that b = ad.

Similarly, since b divides c, by definition,

A2: There is an integer e such that c = be.

Replacing b in A2 with the expression in A1, one has

A3: c = be = ade = a(de).

Looking at B1, the desired value for the integer f is de.

Condensed Proof: Since a divides b, by definition, there is an integer d such that b = ad. Similarly, since b divides c, there is an integer e such that c = be. But then c = be = a(de). On setting f = de, it follows by definition that a divides c.

5.3 (1) Condensed Proof: Assume, to the contrary, that everyone has a different number of hairs on their heads. Number the people so that the person with the fewest hairs receives the number 1, the person with the next fewest hairs receives the number 2, and so on. Since everyone has a different number of hairs, the person whose number is 2 has at least 1 hair, and the person whose number is 1,000,001 has at least 1,000,000. But no one can have more than 250,000 hairs, and so a contradiction has been reached.

(2) Condensed Proof: Assume, to the contrary, that everyone has a different number of friends. Number the people so that the person with number 1 has the fewest friends, the person with number 2 has the next fewest friends, and so on. Since everyone has a different number of friends, the person whose number is 1 has no friends, ..., the person whose number is n has n - 1 friends. In other words, this last person is friends with everyone, including the person who has no friends, a contradiction.

(3) Either 0 or more than one.

CHAPTER 6

6.1 (1) There is a unique line through two distinct points.

(2) There is a unique line through a point P not on a line L that is perpendicular to L.

(3) In a right triangle, there is a unique right angle.

(4) If a, b, c, d, e, and f are real numbers with ad - bc $\neq$ 0, then there is a unique solution to the system of equations ax + by = e and cx + dy = f.

(5) Not unique.

(6) Unique.

(7) Unique.

(8) Not unique.

(9) Not unique.

6.2 (1) Certain Property: $y < 0$

Something Happens: $x = (2 - x)y$

A1: $y < 0$ and $x = (2 - x)y$

A2: $z < 0$ and $x = (2 - x)z$

B1: $y = z$

$2 - x$.

(2) The proof has omitted the discussion of existence. Only the issue of uniqueness was addressed.

(3) Analysis: Recognizing the key word "unique" in the conclusion, the direct uniqueness method will be used. First, an $\angle Y$ must be constructed so that $\angle X$ and $\angle Y$ are complementary. But, by hypothesis, $\angle X$ is acute so, be definition,

A1: $\angle X < 90°$.

From A1, it is possible to construct an $\angle Y$ whose measure is $90° - \angle X$. It is not hard to see that $\angle X$ and $\angle Y$ are complementary because $\angle X + \angle Y = 90°$. To show uniqueness, suppose that

A2: $\angle Y$ is complementary to $\angle X$.

A3: $\angle Z$ is complementary to $\angle X$.

It must be shown that

B1: $\angle Y = \angle Z$.

By the definition of complementary angles, it follows from A2 and A3 that

A4: $\angle X + \angle Y = 90°$.

A5: $\angle X + \angle Z = 90°$.

Combining A4 and A5 one has

A6: $\angle X + \angle Y = \angle X + \angle Z$.

On subtracting $\angle X$ from both sides of A6 one obtains precisely B1, and so the uniqueness is established.

Condensed Proof: To construct an $\angle Y$ that is complementary to $\angle X$, note that $\angle X < 90°$ because $\angle X$ is acute. So construct an $\angle Y$ whose measure is $90° - \angle X$. Then $\angle X + \angle Y = 90°$ and so the two angles would be complementary. Suppose now that $\angle Z$ is also complementary to $\angle X$, so then, $\angle X + \angle Z = 90°$. But then, $\angle X + \angle Y = \angle X + \angle Z = 90°$, from which it follows that $\angle Y = \angle Z$, thus completing the proof.

(4) Analysis: Recognizing the key word "unique" in the conclusion, the direct uniqueness method will be used. First, one must construct x with $x > 0$ such that $xy = 2$. Working forward from the hypothesis that $y > 0$, it is possible to construct $x = 2/y$. Since $2 > 0$ and $y > 0$, $x = 2/y$ is greater than 0. Also, $xy = (2/y)y = 2$. Thus the construction of x is correct. To show uniqueness, suppose that

A1: $x > 0$ and $xy = 2$.

A2: $z > 0$ and $zy = 2$.

It must be shown that

B1: $x = z$.

From A1 and A2 one has

A3: $xy = zy$.

Since $y > 0$, one can divide both sides of A3 by y to obtain precisely B1, thus completing the proof.

Condensed Proof: Construct $x = 2/y$, which can be done since $y > 0$. Clearly, the numerator 2 and the denominator y are both positive, so $x > 0$. Also $xy = 2$. Now let z be another positive number satisfying $zy = 2$. But then $xy = zy$, and on dividing by the positive number y one obtains $x = z$, completing the proof.

(5) Analysis: Recognizing the key word "unique" in the conclusion, the direct uniqueness method will be used. First, it is necessary to construct the desired x and y. To do so, one can multiply the first equation by d and the bottom one by b. On subtracting the resulting two equations one obtains $(ad - bc)x = ed - bf$. Using the hypothesis that $ad - bc \neq 0$, one can divide both sides by $ad - bc$ to obtain $x = (ed - bf)/(ad - bc)$. A similar set of algebraic operations produces $y = (af - ce)/(ad - bc)$. To show uniqueness, suppose that (x_1, y_1) and (x_2, y_2) both satisfy

A1: (1) $ax_1 + by_1 = e$ and (2) $cx_1 + dy_1 = f$.

A2: (3) $ax_2 + by_2 = e$ and (4) $cx_2 + dy_2 = f$.

It must be shown that

B1: $(x_1, y_1) = (x_2, y_2)$.

Working backward from B1, one can ask the Key Question: How can I show that two pairs of real numbers (namely (x_1, y_1) and (x_2, y_2)) are equal?

Using Definition 11, it must be shown that

B2: $x_1 = x_2$ and $y_1 = y_2$.

To see that $x_1 = x_2$, subtract (3) in A2 from (1) in A1 to obtain

A3: $a(x_1 - x_2) + b(y_1 - y_2) = 0$.

Then subtract (4) in A2 from (2) in A1 to obtain

A4: $c(x_1 - x_2) + d(y_1 - y_2) = 0$.

Now multiply A3 by d and A4 by b and subtract the resulting equations to obtain

A5: $(ad - bc)(x_1 - x_2) = 0$.

Finally, use the hypothesis that ad - bc $\neq$ 0 to divide both sides of A5 by ad - bc to obtain

A6: $x_1 - x_2 = 0$, or, $x_1 = x_2$.

A similar set of operations yields $y_1 = y_2$, and thus B2 has been established, completing the proof.

Condensed Proof: Use the fact that ad - bc $\neq$ 0 to construct x = (ed - bf)/(ad - bc) and y = (af - ce)/(ad - bc). It is easy to see that ax + by = e and cx + dy = f. Now suppose that (x_1, y_1) and (x_2, y_2) satisfy (1) $ax_1 + by_1 = e$, (2) $cx_1 + dy_1 = f$, (3) $ax_2 + by_2 = e$, and (4) $cx_2 + dy_2 = f$. Subtracting (3) from (1) and (4) from (2) yields (5) $a(x_1 - x_2) + b(y_1 - y_2) = 0$ and (6) $c(x_1 - x_2) + d(y_1 - y_2) = 0$. Multiplying (5) by d and (6) by b and subtracting yields $(ad - bc)(x_1 - x_2) = 0$, and so $x_1 = x_2$. Similarly, $y_1 = y_2$ so $(x_1, y_1) = (x_2, y_2)$.

(6) Analysis: Recognizing the key word "unique" in the conclusion, the direct uniqueness method will be used. Accordingly, one must first construct the complex number (c + di) such that (a + bi)(c + di) = (ac - bd) + (bc + ad)i = 1. Thus, one needs ac - bd = 1 and bc + ad = 0. Noting from the hypothesis that $a^2 + b^2 \neq 0$, construct $c = a/(a^2 + b^2)$ and $d = -b/(a^2 + b^2)$. It is easy to verify that ac - bd = 1 and bc + ad = 0. To show uniqueness, assume that (e + fi) also satisfies

A1: (a + bi)(e + fi) = (ae - bf) + (be + af)i = 1.

It must be shown that

B1: c + di = e + fi

Working backward from B1, one can ask the
Key Question: How can I show that two complex numbers (namely c + di and e + fi) are equal?

One answer is to show that

B2: Both (c + di) and (e + fi) are equal to a third complex number.

Working forward by multiplying both sides of A1 by c + di and using the rules of complex multiplication, one has:

A2: (c + di)[(a + bi)(e + fi)] = [(c + di)(a + bi)](e + fi) = (1)(e + fi) = e + fi.

Also, since (a + bi)(e + fi) = 1 (See A1),

A3: (c + di)[(a + bi)(e + fi)] = (c + di)(1) = c + di.

Thus, both (c + di) and (e + fi) are equal to (c + di)[(a + bi) (e + fi)], and so B2 has been established, thus completing the proof.

Condensed Proof: Since a and b are not both 0, $a^2 + b^2 \neq 0$ and so one can construct $c = a/(a^2 + b^2)$ and $d = -b/(a^2 + b^2)$. It is easy to verify that $(a + bi)(c + di) = 1$. Now suppose that $(e + fi)$ also satisfies $(a + bi)(e + fi) = 1$. Then $(e + fi)(1) = (e + fi)[(a + bi)(c + di)] = [(e + fi)(a + bi)](c + di) = (1)(c + di) = c + di$, thus completing the proof.

6.3 (1) Analysis: Since the conclusion contains the key word "unique," the indirect uniqueness method will be used. The fact that the line crosses the x-axis was established in Example 9. To establish the uniqueness by the indirect method, assume that

A1: $0 = mx + b$.

A2: $0 = mw + b$.

A3: $x \neq w$.

A contradiction to the hypothesis that $m \neq 0$ will be reached by working forward from A1, A2, and A3 to show that $m = 0$. Specifically, from A1 and A2,

A4: $mx + b = mw + b$.

On subtracting b from both sides one has

A5: $mx = mw$, or, $m(x - w) = 0$.

Using A3, one can divide both sides of A5 by $(x - w)$ to obtain the contradiction that $m = 0$.

Condensed Proof: An x such that $0 = mx + b$ was already constructed in Example 9. To show the uniqueness, suppose that w also satisfies $0 = mw + b$ and that $w \neq x$. But then, $mx + b = mw + b$, or $mx = mw$, or $m(x - w) = 0$. Since $x \neq w$, it follows that $m = 0$, contradicting the hypothesis that $m \neq 0$.

(2) Analysis: Recognizing the key word "unique" in the conclusion, the indirect uniqueness method will be used. First one must construct an angle Y that is supplementary to $\angle X$. Since, by hypothesis, $\angle X < 180°$; one can construct an $\angle Y$ whose measure is $180° - \angle X$. It is easy to see that $\angle X$ and $\angle Y$ are supplementary because $\angle X + \angle Y = 180°$. To show the uniqueness, suppose that

A1: $\angle Y$ is supplementary to $\angle X$.

A2: $\angle Z$ is supplementary to $\angle X$.

A3: $\angle Y \neq \angle Z$ (in terms of their measure).

Working forward from A1 and A2 by definition, one has

A4: $\angle X + \angle Y = 180°$.

A5: $\angle X + \angle Z = 180°$.

Combining A4 and A5 one has that

A6: $\angle X + \angle Y = \angle X + \angle Z$.

However, on adding $\angle X$ to both sides of A3 one obtains

A7: $\angle X + \angle Y \neq \angle X + \angle Z$.

Clearly A6 and A7 are contradictory.

Condensed Proof: To construct $\angle Y$, noting that $\angle X < 180°$, set $\angle Y = 180° - \angle X$. Then $\angle X$ and $\angle Y$ are supplementary because $\angle X + \angle Y = 180°$. Now suppose that $\angle Z$ is also supplementary to $\angle X$ but that $\angle Y \neq \angle Z$. Then $180° = \angle X + \angle Y = \angle X + \angle Z$. But on the other hand, since $\angle Y \neq \angle Z$, $\angle X + \angle Y \neq \angle X + \angle Z$, and this contradiction establishes the claim.

(3) Analysis: Recognizing the key word "unique" in the conclusion, the indirect uniqueness method will be used. First one must construct the desired integer x. Clearly $x = 5$ satisfies $x > 0$ and $x^2 = 25$. To establish the uniqueness, suppose that

A1: $x > 0$ and $x^2 = 25$.

A2: $y > 0$ and $y^2 = 25$.

A3: $x \neq y$.

Working forward from A1 and A2 it follows that

A4: $x^2 = y^2$, or, $x^2 - y^2 = 0$.

By factoring, A4 can be rewritten as

A5: $(x - y)(x + y) = 0$.

From A3 it follows that $(x - y) \neq 0$ and so one can divide both sides of A5 by $(x - y)$ to obtain

A6: $x + y = 0$, or, $x = -y$.

From A2, y is greater than 0, so -y is less than 0. Thus, from A6,

A7: $x < 0$.

A7 contradicts the fact that $x > 0$ in A1, completing the proof.

Condensed Proof: Clearly, the value of 5 for x satisfies $x > 0$ and $x^2 = 25$. Now suppose that y also satisfies $y > 0$ and $y^2 = 25$, but $y \neq x$. Then it follows that $x^2 = y^2$, or $(x - y)(x + y) = 0$. Since $x \neq y$, one can divide by $(x - y)$ to obtain $x + y = 0$, or $x = -y$. Since $y > 0$, $-y < 0$ and thus $x < 0$, which contradicts the fact that $x > 0$.

(4) Analysis: Recognizing the key word "unique" in the conclusion, the indirect uniqueness method will be used. Thus one must first show that a right triangle has a right angle. This fact, of course, is true by the definition of a right triangle. To establish the uniqueness, suppose that

A1: $\angle X$ is a right angle of a right triangle.

A2: $\angle Y$ is a right angle of a right triangle.

A3: $\angle X$ is different from $\angle Y$.

Working forward by definition from A1 and A2,

A4: $\angle X = 90°$.

A5: $\angle Y = 90°$.

A contradiction is obtained by adding up the three angles of the right triangle. So let $\angle Z$ be the third angle. Then

A6: $\angle X + \angle Y + \angle Z = 90° + 90° + \angle Z = 180° + \angle Z$.

Since $\angle Z > 0°$, A6 states that the three angles of the triangle add up to more than 180°, and this contradiction completes the proof.

Condensed Proof: Clearly, a right triangle has a right angle by definition. Suppose that $\angle X$ and $\angle Y$ are two right angles, so $\angle X = \angle Y = 90°$. But then, if $\angle Z$ is the third angle of the triangle, then $\angle X + \angle Y + \angle Z = 180° + \angle Z$. Since $\angle Z > 0°$, the angles of this triangle would add up to more than 180°, a contradiction.

(5) Analysis: Recognizing the key word "unique" in the conclusion, the indirect uniqueness method will be used. However, the issue of existence will be omitted. To establish the uniqueness, suppose that x and y are real numbers satisfying

A1: $x^3 = r$.

A2: $y^3 = r$.

A3: $x \neq y$.

Working forward from A1, A2, and A3, it will be shown that $r = 0$, thus contradicting the hypothesis that $r > 0$. Specifically, from A1 and A2 it follows that

A4: $x^3 = y^3$, or, $x^3 - y^3 = 0$.

On factoring A4 one has

A5: $(x - y)(x^2 + xy + y^2) = 0$.

Using the fact that $x \neq y$ from A3, it is possible to divide both sides of A5 by $(x - y)$ to obtain

A6: $x^2 + xy + y^2 = 0$.

Thinking of A6 as a quadratic equation of the form $ax^2 + bx + c = 0$ where $a = 1$, $b = y$, and $c = y^2$, the quadratic formula yields

A7: $x = (-y \pm \sqrt{y^2 - 4y^2})/2 = (-y \pm \sqrt{-3y^2})/2$.

Since x is real, and the expression for x in A7 requires taking the square root of $-3y^2$, it must be that

A8: $y = 0$.

But then from A2 one has that $r = y^3 = 0$, and this contradicts the hypothesis that $r > 0$.

Condensed Proof: Only the issue of uniqueness will be addressed, so assume that $x \neq y$ and $x^3 = y^3 = r$. Then $x^3 - y^3 = (x - y)(x^2 + xy + y^2) = 0$. Since $x \neq y$, $x^2 + xy + y^2 = 0$ and, by the quadratic formula, $x = (-y \pm \sqrt{-3y^2})/2$. Since x is a real number, y must be 0, but then, $r = y^3 = 0$, contradicting the hypothesis that $r > 0$.

CHAPTER 7

7.1 (1) Forward: n is an odd integer.

Backward: n^2 is an odd integer.

(2) Forward: $f(x) = f(y)$.

Backward: $x = y$.

(3) Forward: The quadrilateral RSTU is not a rectangle.

Backward: Some angle of the quadrilateral RSTU is obtuse.

(4) Forward: The equation $n^2 + n = c$ has an odd integer solution for n.

Backward: c is an even integer.

(5) (d)

(6) (b)

(7) Analysis: Recognizing the key word "not" in the conclusion ($\neq$), the contrapositive or contradiction method could be used. Accordingly, it can be assumed that

NOT B: $mx + b = mw + b$.

It must be shown that

NOT A: $x = w$.

To do so, work forward from NOT B by subtracting b from both sides to obtain

A1: $mx = mw$.

Using the hypothesis that $m \neq 0$, one can divide both sides of A1 by m to obtain precisely NOT A, thus completing the proof.

Condensed Proof: Assume, to the contrary, that $mx + b = mw + b$. But then, $mx = mw$, and since $m \neq 0$ by hypothesis, it follows that $x = w$, completing the proof.

(8) Analysis: Recognizing the key word "no" in the conclusion, the contrapositive or contradiction method can be used. Here, the contrapositive method will be used. Accordingly, it can be assumed that

NOT B: There is an odd integer n such that $n^2 + n = c$.

It must be shown that

NOT A: c is an even integer.

Working backward from NOT A, one is led to the
Key Question: How can I show that an integer is even?

Using a definition, the new statement to be proved is:

B1: There is an integer k such that $c = 2k$.

Recognizing the key words "there is" in B1, the construction method will be used to produce the value for k. Turning to the forward process, in NOT B it says that n is odd so, by definition, one has

A1: There is an integer p such that $n = 2p + 1$.

Replacing this expression for n in NOT B one has:

A2: $c = n^2 + n = (2p + 1)^2 + (2p + 1) = (4p^2 + 4p + 1) + 2p + 1 =$
$4p^2 + 6p + 2 = 2(2p^2 + 3p + 1)$.

Looking at B1 and at A2, it is not hard to see that the desired value for the integer k is $2p^2 + 3p + 1$.

Condensed Proof: Assume, to the contrary, that there is an odd integer n satisfying $n^2 + n = c$. Consequently, there is an integer p such that $n = 2p + 1$, by the definition of an odd integer. But then, $c = n^2 + n = (2p + 1)^2 + (2p + 1) = 2(2p^2 + 3p + 1)$, and so c would be an even integer, thus completing the proof.

7.2 (1) True;

Converse: If a triangle has two complementary angles, then the triangle is a right triangle; true.

Contrapositive: If a triangle does not have two complementary angles, then the triangle is not a right triangle; true.

Inverse: If a triangle is not a right triangle, then the triangle does not have two complementary angles; true.

(2) False;

Converse: If R is perpendicular to T and S is perpendicular to T, then the lines R and S are parallel; true.

Contrapositive: If either R is not perpendicular to T or S is not perpendicular to T, then R is not parallel to S; false.

Inverse: If R is not parallel to S, then either R is not perpendicular to T or S is not perpendicular to T; true.

(3) False;

Converse: If the quadrilateral WXYZ is a rectangle, then the quadrilateral WXYZ is a parallelogram with at least one right angle; true.

Contrapositive: If the quadrilateral WXYZ is not a rectangle, then the quadrilateral WXYZ is not a parallelogram with at least one right angle; false.

Inverse: If the quadrilateral WXYZ is not a parallelogram with at least one right angle, then the quadrilateral WXYZ is not a rectangle; true.

(4) Statement: NOT A implies NOT B.

A	B	NOT A	NOT B	NOT A => NOT B
T	T	F	F	T
T	F	F	T	T
F	T	T	F	F
F	F	T	T	T

(5) Statement: (NOT A) OR B.

A	B	NOT A	(NOT A) OR B
T	T	F	T
T	F	F	F
F	T	T	T
F	F	T	T

(6) They are identical. Both "A implies B" and (NOT A) OR B are true in all cases except when A is true and B is false.

CHAPTER 8

8.1 (1) Uniqueness method because of the key words "one and only one."

(2) Contrapositive or contradiction method because of the key word "no" in the conclusion.

(3) Forward-Backward method because the conclusion does not contain any recognizable key words.

(4) Contrapositive or contradiction method because of the key word "not" in the conclusion.

(5) Construction method because of the key words "there are" in the conclusion.

(6) Forward-Backward method because the conclusion does not contain any recognizable key words.

GLOSSARY OF MATHEMATICAL SYMBOLS

Symbol	Meaning	Page
=>	implies	7
$\wedge$	and	13
$\vee$	or	14
$\perp$	perpendicular	37
Q.E.D.	quod erat demonstrandum (which was to be demonstrated)	58
<=>	if and only if	107
$\sim$	not	125
$\exists$	there is (there are, etc.)	155
$\ni$	such that	155

INDEX